# The Final Analysis of Weights from Port Royal, Jamaica

C. Wayne Smith

BAR International Series 675
1997

Published in 2019 by
BAR Publishing, Oxford

BAR International Series 675

*The Final Analysis of Weights from Port Royal, Jamaica*



ISBN 9780860549147 paperback
ISBN 9781407349862 e-book

DOI https://doi.org/10.30861/9780860549147

A catalogue record for this book is available from the British Library

This book is available at www.barpublishing.com

BAR Publishing is the trading name of British Archaeological Reports (Oxford) Ltd. British Archaeological Reports was first incorporated in 1974 to publish the BAR Series, International and British. In 1992 Hadrian Books Ltd became part of the BAR group. This volume was originally published by John and Erica Hedges in conjunction with British Archaeological Reports (Oxford) Ltd / Hadrian Books Ltd, the Series principal publisher, in 1997. This present volume is published by BAR Publishing, 2019.

BAR
PUBLISHING

BAR titles are available from:

BAR Publishing
122 Banbury Rd, Oxford, OX2 7BP, UK
EMAIL info@barpublishing.com
PHONE +44 (0)1865 310431
FAX +44 (0)1865 316916
www.barpublishing.com

**Analysis of the Weight Assemblage of Port Royal, Jamaica**

C. Wayne Smith, Ph D.

## ABSTRACT

The assemblage of weights recovered from excavations at Port Royal, Jamaica is the largest collection of seventeenth-century weights recovered from a single colonial site. On June 7, 1692, just shortly before noon, an earthquake shook the community of Port Royal and in a matter of minutes, due to a process known as liquefaction, approximately thirty-two acres of the colonial port community sank below the waters of Kingston Harbor. To date, 90 weights have been recovered from nautical and terrestrial excavations at the site. Many of these bear the stamps and ciphers of English trade guilds as well as owners marks and regal stamps of authority. Several weights in the collection also bear a mark in the shape of a dagger or sword, which is associated with the City of London.

Bronze and lead weights bearing cipher stamps offer a unique opportunity for archaeological and historical investigation. Combining data from archaeological excavations with information from wills, inventories and data from comparative assemblage known as the Streeter Collection, will result in a multi disciplinary analysis of the weights collection that has the potential for contributing a great deal of information about local and long distance trade and commercial patterns in colonial Jamaica. From a micro perspective, this analysis will contribute to the broader understanding of life and economic activities within the excavated area of the colonial port community. From a macro perspective, analysis of the weights should corroborate reforms in English law regarding the economic necessity for standardization of weights. Analysis of weight iconography suggests differences in mind-set that may have contributed in placing England apart from other European countries, in an advantageous position to become a leading economic force in the seventeenth century. An in depth analysis of the weight collection associated with seventeenth-century Port Royal will draw many sources of data together, offering new perspectives on the colonial Jamaican process.

This research is dedicated to Helen Dewolf, who has always been supportive of my efforts. Her quiet strength and support have always been my encouragement to pursue those things that are my passions.

## ACKNOWLEDGMENTS

It is only when you set out to thank all the people who have encouraged you in your academic pursuits, that you realize the importance of their contributions in attaining your own personal goals. I am greatly indebted to Dr. D.L. Hamilton for the privilege of having been able to study and work with him at the Conservation Research Laboratory at Texas A&M University and in the field for so many years. He has generously allowed me to tap his knowledge of historical archaeology and archaeological conservation and it is his excitement for research and subtle encouragement that have prompted me to pursue archaeology and the conservation of archaeological materials as a vocation. His professionalism, friendship and constant expectation for results have been invaluable to me.

I am very grateful to Dr. Grider for her perspectives on material culture studies - her encouragement and belief in the need to go beyond simple artifact descriptions has been a compelling force in my research into the semiotic aspects of weight iconography. Her commitment to the interdisciplinary approach of archaeological research has brought new perspectives and insights to my research. By focusing my attentions on World Systems Theory, Dr. Carlson has given me a solid framework from which to view cultural activities in colonial Jamaica. His perspectives on historical archaeology have been greatly appreciated.

Dr. Schmidt has helped me to see that history and anthropology maintain a comfortable symbiotic relationship. His perspectives on Latin American history have played an important part in shaping my research strategies and his personal enthusiasm for research has been inspirational.

I am also very grateful for the friendship and knowledge that I have gained from Mr. Richard McClure, who has been the curator of artifacts at the Old Naval Hospital in Port Royal, Jamaica. His assistance in photographing artifacts, his extensive personal knowledge and his constant support have made working with him a pleasure. Richard is a true friend who has worked long hours to preserve the history and artifact assemblage of Jamaica. Due to his own quest for knowledge and tireless enthusiasm in maintaining the storehouse of artifacts from archaeological excavations on the island, Jamaica has amassed an impressive material culture assemblage to document the pre and post-colonial periods of Jamaican history.

While finding my way as a undergraduate student, Dr. Michael Spence of the Anthropology Department at the University of Western Ontario encouraged me to stretch my academic wings. Without his influences, I would never have taken tentative steps to pursue a college education. Dr .R.D.Connor of the University of Alberta has spent many hours reviewing the Port Royal catalog of weights and his help in locating essential information has been invaluable. His personal research and text book entitled "Weights and Measures of England" have set the highest standards for research regarding the history of English weights and measures. I am indebted to all of you for the knowledge you have shared - it has been a privilege to have studied with you.

## TABLE OF CONTENTS

## LIST OF FIGURES

**LIST OF TABLES**

# CHAPTER I
# INTRODUCTION -
# HISTORICAL BACKGROUND

The archaeological assemblage of weights recovered from excavations at Port Royal, Jamaica, is the largest collection of seventeenth-century weights recovered from a single colonial site. On June 7, 1692, just shortly before noon, an earthquake shook the town of Port Royal and in a matter of minutes, some thirty-two acres of land sank below the waters of Kingston Harbor, due to a process known as liquefaction. From the rubble of numerous buildings, 90 weights in a large variety of types have been recovered. Many of these weights bear the stamps and ciphers of English guilds, monarchs, owners marks and cities of origin.

The fact that Port Royal is a catastrophic site is important. Artifacts recovered from excavations by Robert Marx and the ten years of excavation by Dr. D.L. Hamilton are impressive, both in their size and in the complete nature of many of the artifacts. Nautical excavations at Port Royal have provided a unique collection of weights that would not have survived in most archaeological contexts.

Most of the weights have survived in good shape. Layers of silt and a protective layer of staghorn coral have encased the site, protecting the 1692 strata from both environmental and human intrusion. Although lead weights are relatively soft, many of them bear good diagnostic marks. The use of these diagnostic features in combination with data from the Act of Common Council records has enabled us to accurately associate weights in the assemblage to specific time period as well a specific mercantile activities.

Weights tend to become invisible in many archaeological contexts. Often, lead weights do not appear in the archaeological record because of their important re-use value. Old weights, surplus weights and case window caming fragments were valuable as a material because they could be melted down for the making of musket balls and other armament. Old weights, bearing official ciphers, were often brought to a current standard by a variety of means. The mass of a weight could be adjusted in a variety of ways. Lead plugs or iron nails hammered into the body of a weight were ways to increase mass. Weight adjusters were often nailed onto weights as a means of bringing old weights to a new standard. For a variety of reasons, lead and bronze weights often do not appear in the archaeological record in the numbers that a survey of the archival records suggests should be present. In this respect then, the Port Royal site is an exception to the rule. In a very short period of time, the earthquake sealed a cross section of cultural materials, including weights, into a silt layer, preserving accurately both the quantitative and qualitative nature of the assemblage.

Weights are important diagnostic tools because the process of adapting them to a differing standard, use-wear and the marks and ciphers that they bear all tell something about change through time. Many weights also bear owner's marks which can be compared to the wills and inventories for the town. Many of the bronze disc weights also bear the mark of a monarch, represented as an initial beneath a crown or an orb. While such markings can be useful in establishing an ante-quem date for a site, the historical record for Port Royal is such that a firm date of 1692 can be fixed to archaeological materials that are found in close context with the foundations of all the submerged structures.

The objectives of this research are as follows:
1) To create a descriptive catalog of the weight assemblage. Such a catalog is necessary for analysis of the weight assemblage and will be helpful for future work and curation.
2) Relate the weight assemblage of Port Royal to the Streeter Collection[1] of weights and archival materials in an effort to understand the social and commercial significance of this large a collection of weights.
3) Study the patterns of distribution of weights and other archaeological materials in order to better understand the functions and uses of the excavated buildings.

To understand the nature of the weight assemblage and the frequency with which weights, scales and "stillyards" appear in the inventories, one must delve into commerce and trade in Medieval Europe.

---

[1]The Streeter Collection of weights is permanently housed in the Medical Library at Yale University, New Haven, Connecticut. This collection has been invaluable for comparative studies since it includes a large number of examples of seventeenth-century weights that are included in the Port Royal assemblage.

## CHAPTER II

## HISTORICAL BACKGROUND

### Development and Perception of Weights and Measures

The imagery of justice and equitable process under the law associated with scales has its origins in the distant past. The *Ascension of Isaias*, dated circa 90 AD, referred to the fact that the "great angel Michael always interceded for the human race" with his scales of justice (Thurston & Attwater 1962: 677). Icons from the Egyptian Book of the Dead and the Mendoza Codex of Mexico suggest that systems of measurement span the entire history of social interaction. In his research of the Mexican codices, Antonio y Cama found references to a special system of weights used for the collection of tributes and taxes for the Aztec nobility. (Barlow 1949: 9) It is noteworthy that there seems to have been a parallel development of weight systems throughout the world and in most cases, these weights were developed for the purpose of collecting taxes. Whatever their origins, studies of weight and measurement systems and iconographic references do not in themselves elucidate the true developmental history of weights. The English language is filled with aphorisms that point to long-standing beliefs and understandings that we have kept alive in our everyday conversations. Proverbs XI, 1 of the King James version of the Bible which is dated circa 350 BC, sets high metrological standards in stating that "a false balance is an abomination to the Lord: but a just weight is his delight." Whereas this statement is directed towards encouraging accurate weights and measures, many aphorisms offer words of wisdom for living well. "An ounce of prevention is worth a pound of cure" and "when many strike an anvil they will strike by measure" both encourage introspection and observation as keys to better living (Menchen 1942: 774). Phrases such as the English expression "penny wise and pound foolish" and an old Negro proverb, "don't measure yourself; it will make you die" offer us glimpses at long-held wisdom and the social systems which helped form them. Regardless of cultural background, human language seems to be filled with aphorisms related to weights and measures.

The development of any weight system is the result of a gradual evolutionary process. Social intercourse, the forum in which weight systems are developed, is a complex process which is mediated by innumerable contributing influences. Cultural interactions including the decrees of monarchs, trade, fraud, technological advancements and wars all play an important part in the formation and use of a system of weights. In the case of England's metrological systems, these factors in combination with Britain's early emergence as a center for cultural interaction and international trade activities, make it difficult to trace any singular course or logical developmental process for the development of weights and measures. Indeed, while they are often combined as a singular topic for investigation, weights and measures took developmentally divergent paths at an early stage in English history. The archaeological record indicates that the basic principles used to create scales for the measurement of mass have progressed in a strikingly similar fashion amongst socially isolated populations, in Europe and the New World. The image of a simple scale on a Greek Taleides amphora, dated at the 6th century BC., is similar to scales of the same time period, found in Central America (Sokeland 1900: 553). In many cultures, the custody of ancient weight standards was usually associated with important religious and secular institutions. Egyptian standards were kept in the dromos of the god Anubis at Memphis and the safekeeping of the Hebrew "shekel of the sanctuary" was entrusted to the temple priests. The standards of Athens were closely guarded in the temple of Athena while the Roman metrological standards were kept in the temple of Hercules. During the campaign of Charlemagne in France, all standards were stored in the royal palace and from the early days of English history, Westminster chapel has served as a repository for all official standards. Despite the seeming universal association of weights with the highest offices of church and state in European and Mediterranean societies, only the Anglo-Saxon world attempted to maintain and officially sanction two or more dissimilar metrological systems simultaneously. Historically, many nations utilized foreign systems of measurement as a means to encourage long distance trade and mercantile activity. The English however, specialized their metrological system to a greater extent, and in doing so, created a set of standards that which positioned them an an advantageous position for inter-regional trade. The Troy standard was reserved for the measurement of jewels, precious metals, liquor, bread and in rare cases, pharmaceuticals. The Avoirdupois standard and the Tower pound were used for the measurement of base metals, wool, tallow, hemp and general food items.

Occupation-related technologies were also an important cornerstone in the development of scales, beams and weight standards. Throughout her history, bread has always been an important staple commodity in England. Wastel or common white bread became the standard in terms of production processes, price and weight by which all other bread types were produced and priced. A drawing from Beni-Hassan shows an ancient Egyptian goldsmith using an equal-arm scale to measure his gold (Sokeland 1900: 553). Goldsmiths throughout history have utilized a special standard other than the standard used for common commodities. While Egyptians did not invent the equal-arm scale, they refined the mechanical designs of the device, improving its ability to accurately measure very small commodities.

Hieroglyphs show that the free-moving arms of the earliest Egyptian balances were supported from below. Many of the early improvements to scales, such as the plumb bob leveling system, were simple adaptations. The plumb bob indicator, which was a weight supported by three strings, was an ingenious concept that took into account three dimensional positioningas a factor inconsidering the level position for a scale arm. Only when all three strings were taut was the balance arm in a perfectly horizontal resting position. This position indicated a proper measurement. Any deviation from the horizontal position meant that one or more of the strings remained slack. Besides the plumb bob system, early technological improvements included the use of an indicator tongue and silk threads to support scale pans. Figure 1 is an Egyptian rendering showing a scale in use. In the scale pan located at the load end of the arm is a commodity which is being weighed. A zoomorphic weight shaped in the form of a bull's head is positioned in the opposite scale pan. The merchant is depicted as inspecting the three strings of the plumb bob. When these strings are hanging taut, he will have accurately weighed his commodity.

The ancient Egyptian in Figure 1 is using an early period scale system in which the balance arm was supported from below. By the height of the Egyptian Empire, scales were suspended from a pivot point as a means of improving their performance and accuracy. The processes of refinement and adaption were slow but by the end of the Egyptian Empire, the accuracy of scales had been improved greatly, resulting in scales and balance beams that have remained largely unchanged to date. Refinements to the suspension system of the scale arm and a sharply defined fixed pivotal point (commonly called a knife edge) further increased the accuracy of scales. With these innovations, scales and beams could be used for a greater range of commercial activities. All of these developments were aimed at increasing the accuracy of scales, which indicates that early Egyptians perceived a need for more accurate scale and weight systems and utilized their technology to develop better metrological systems (Sokeland 1900: 554).

**Figure 1** An Egyptian merchant inspecting the plumb bob of his scale (Sokeland 1900:555

Dessemers or unequal-armed measuring beams (also known as steelyards) were popular for commercial exchanges throughout Europe. Widely used in early Rome, and commonly known as the Roman scale, dessemers were manufactured in a wide range of sizes capable of measuring commodities such as iron, as well as smaller, valuable commodities such as spices. These devices operated on a lever principle with one arm of the pair being much longer than the other. Each dessemer had a single, hollow counterpoise or moveable counter weight which was moved along the long arm of the beam until the arm rested in a horizontal position. The weight of the commodity was determined by reading the scale markings which were etched along the length of the beam arm. Many historic documents report that merchants preferred dessemers to more simply designed balance scales because they were easier to manipulate for fraudulent dealings with unsuspecting customers.

Although the origin of the dessemer is unclear, the Old Testament makes several references to the extremely common use of false beams, weights and measures associated with mercantile exchange. Judaic and Islamic laws both condemn having two differing weight systems in the same house, which suggests that fraudulent exchanges were the widespread rule and not the exception (Brewer 1853: 314). The problem of double standards in weight systems was not isolated to commercial activities in England. Witold Kula has noted that in Poland, as elsewhere in Medieval Europe, grain was customarily sold in heaped bushels. Heaped measurements were problematic because no standards had been established regarding the size and shape of the vessels used to measure grain (Kula 1986: 55, Pasley 1834: 93). If a merchant used a larger circumference vessel to purchase grain, the capacity for heaping the measure was greatly increased. By using a vessel with a smaller circumference to sell grains, the potential for heaping the measure was greatly reduced. The difference in diameter between the two vessel types and the resulting difference in heaped capacity was an important source of fraudulent profit potential for the merchant (Kula 1986: 55, Pasley 1834: 92). Because this practice was widespread throughout Europe, laws regarding the capacity and physical size of vessels were rapidly enacted as well as legislation calling for the striking of the measure. Striking the measure simply meant that a flat rod was drawn across the filled vessel before the load was weighed. Removal of the heaped proportion of grain eliminated some of the problem of variance in measures of capacity and diminished the margin for variance that served as a merchants added profit margin. But unlike fraudulent containers which were easily detected, variance between weights was more difficult to determine, since each merchant used his weights on a scale that was calibrated to indicate that the weights were correct. Since numerous shapes, sizes and materials were used in the production of weights, the road leading to their standardization was considerably more difficult than the relatively straight path of development for capacity measurement.

The earliest weights recovered from archaeological excavations are made of glass and stone. Unlike iron, which is affected by a wide range of environmental factors that can alter the weight and physical stability of a object, these materials were largely non-corrosive (Hamilton 1976: 8). They were perceived as desirable materials for making weights since they appeared to retain their size, weight and overall dimensional form over a long period of time. Excellent examples of glass and stone weights have been associated with Babylonian, Assyrian, Egyptian, Judean, Indian and Italian cultures. Glass weights and zoomorphic weights have been recovered in large quantities from numerous Near Eastern sites including the Ulu Burun nautical excavations, conducted by Texas A&M University's Institute of Nautical Archaeology, in Turkey. Vessels such as those found at Ulu Burun and Serçe Liman probably transported goods and materials between ports around the Mediterranean, necessitating the use of several weight systems to facilitate commercial transactions. The large quantity of glass, stone and bronze weights recovered from nautical and terrestrial sites suggests that ancient sea-going merchant-traders were well versed in the use of and conversion between different weight systems. There can be no doubt that these mariners were cognizant of the need for stable, standardized weights.

The wide-spread use of glass and serpentine money weights was popular in Islamic countries well into Medieval times (Kisch 1975: 80). These weights were popular because they were perceived to be more accurate than metal weights as well as more difficult to falsify. During the height of Imperial Rome, Byzantine and Roman merchants made the transition to bronze weights, which as their sheer numbers in the archaeological record might suggest, were preferred to the stone and glass weights of their predecessors. A broad survey of European weight collections indicates that the practice of combining metals for the manufacture of weights was more prevalent in Northern Europe. It is probable that the tradition of making weights from compound materials originated in

Scandinavia. These weights and measures were influential in England and Europe, acting as a forerunner to the mark standard which was widespread in its popularity until the Troy system was adopted (Nightingale 1985: 194). Viking weights, included as funerary offerings, were commonly manufactured by sheathing a core of lead or iron with brass (Kisch 1975: 83). By the 8th and 9th century, however, Anglo-Saxon weight makers appear to have adopted this practice and were producing lead core weights with ornately decorated brass casings.

The use of brass for weight manufacturing spread rapidly throughout Europe after the Roman invasion of central Europe in 47 AD, and England appears to have been a focal point for interacting technologies. Similar to the merchant class of the Iberian peninsula, English merchants were at the cultural and commercial crossroads between Roman, Scandinavian, and Moorish influences. German and Saxon standards for weights and measures which had survived the Norman conquest contributed greatly to the formation of a distinctly English system of weights and measures. In spite of this easy cultural and commercial interaction, the Celts maintained a system of weights and measures that were also influential in the formation of the earliest English metrological system. Numismatic and documented evidence suggests that there were at least four different weight standards for the English monetary pound between the time of Æthelred II (AD 1018) and the Norman conquest up to AD 1066 (Nightingale 1985: 195). With such diverse cultural influences, tracing a single origin for early Anglo-Saxon weights is not an easy task and no written form of Saxon assizes[1] has survived to aid in answering the questions of origins and influences that converged in pre-Medieval England to form a unique metrological systems (Connor 1987: 194).

The earliest Anglo-Saxon weights and measures associated with Winchester Cathedral date from the reign of King Edgar (958-975). By the 9th century, political factions in England had begun to unite, and the Saxon measures of Winchester were adopted as the standard for the realm. Between 1017 and 1095, there was a great deal of cultural parlaying which affected the English standards. In 1017, Cnut invaded and conquered England and replaced Æthelred's fifteen-ounce pound with a ten ounce pound. In 1026 however, he altered this standard from 24.5 grams to 27 grams so that it matched the weight of a Roman ounce, which was popular throughout the Byzantine Empire. At the end of the tenth century, Muslim dinars were freely circulating in Iberia and the Catalans struck a new silver currency that was in accord with the Muslim silver dirhem as a means of unifying their gold and silver systems (Nightingale 1985: 197-8).

Recognizing that controlling a unified system of weights and measures was key to regulating the English economy, William the Conqueror decreed that all weights throughout the realm were to be uniform in standard and marked with his seal. As the first step of unification, he ordered that the standards for weights and the trial plates for gold and silver coins were to be moved to London and preserved in the crypt of Edward the Confessor at Westminster Abbey (Hallock 1906; 30). Over the next sixty years, the system of monarchical control over monetary and metrological systems expanded and Edward's crypt reflects this expanded state control since it was enlarged as a repository and renamed Pyx Chapel.

During the reign of King John (1199-1216), the first formal legislation aimed at standardization was enacted. Chapter thirty-five of the Magna Carta of 1215 stated that special standards for wine, ale and corn were to be employed throughout the realm (Connor 1987: 150, Hallock 1906: 32). Although much of the wording of the 1215 Magna Carta was repeated verbatim in Henry III"s act of 1225 and almost identically in Edward I's act of 1297, it appears that the preoccupation of monarchs with unification of the standards was at odds with mercantile activities of shire towns and nobles. Many of these were unwilling to relinquish their grip on alternative systems of weights and measures which deviated from the king's standard.

The importance of the Magna Carta is bound up in the fact that, in many respects, it was the first blueprint for a monarchically dictated English economic system. Chapter Thirty-Five of the document included legislation designed

1

The Houghton Mifflin Canadian Dictionary of the English Language describes assizes as decrees or edicts rendered during legislative sessions for the regulation of weights and measures.

for sweeping reforms in the laws governing all aspects of mercantile activities, including weights and measures. Special provisions were also outlined for a standard breadth for three types of cloth. While the Magna Carta did acknowledge the need for uniformity of measurement, the monetary standard known as the pound, which was first noted in this document, was never defined in legislation until the reign of Henry III in 1266. As Stuart Peterfreund has noted, the Magna Carta denied free evolution of social structures because it formulated government-designed social controls that pervaded all aspects of English life. Through the guild system, the Crown had explicit control over all standards and maintained the right to inspect, correct and levy fines as it saw fit. The Magna Carta also established the freemen of the City (London) as being socially superior to all non residents. Establishing the London standards as superior to all others, businessmen and international traders were forced to adhere to the English system which, from its inception, had been largely established to protect the interests of all freemen of the City (Peterfreund 1989: 40-2). Apart from a long list of sweeping reforms, the Magna Carta set English standards as the frame of reference against which other European standards were measured. Traders and merchants wishing to conduct business with Englishmen had to do so using the English system.

During the same year, the Assize of Bread and Ale was issued. This document set new standards for regulating the weight and prices for the sale of these essential items. Strict fines were levied against bakers and brewers who violated the regulations of the assize. If a baker's bread, commonly sold as a farthing loaf, was lighter than the standard by 2 shillings or less, "then he be amerced" (fined) but if his deviation from the standard exceeded 2 shillings in weight, he was sent to the pillory without an option of being able to redeem himself with the buyer with a cash compensation (Connor 1987: 198). Beer, ale and wine were considerably important in the English diet during the Middle Ages. The nutritional value of bread as a staple food, however, made its regulation, production and sale of paramount importance. Bread manufactured in all major centers was divided into six basic categories, depending on the type and amount of processing required to produce a finished loaf. Table 1 is a brief list of the main bread types and the process associated with the production of each. It is important to note that the production of wastel bread was very tightly regulated because apart from being the common bread for daily consumption, it also served as the standard for setting the price of bread in subsequent assizes after 1215 AD. The tight control over the production of wastel bread and its association with the Troy standard is present as a common theme in Medieval art.

Table 1 Common types of bread

| Bread Types | Process and Use |
|---|---|
| Wastel / Wholemeal | basic white bread |
| Cocket | second quality bread |
| Simnel | better quality bread |
| Wholemeal / Wheaten | made from whole grain with bran removed |
| Treet | bread made from coarse flour |
| Common Wheat | common wheat bread / mixed grains |

The Assize of Bread and Ale of 1266 is unique because it equated a quantity of bread or ale (measured by weight) with a monetary value. Ale was the common beverage of the masses in London because much of the water supply was polluted due to privies draining directly into the local water supply. This value equivalent with the monetary standard allowed local officials, merchants and citizens to more easily adhere to the standard. The Assize also instituted the first use of the Sterling, which is an English penny. Each Sterling was to weigh the equivalent of thirty-two "wheatcorns" taken from the middle of the ear of grain. As it appears in the Coventry Leeds Book for 1474 the definition of a Sterling was explained as "XXII graynes of wheat out of the mydens of the Ere maketh a sterling otherwise called a penny and XX sterlings maketh an ounce and XII ounces maketh a Pounde for sylver, golde, bred and measure" (Connor 1987: 197). Based on this standard, the relationship between the Sterling, pence, pounds, gallons and bushels was established and is illustrated in Table 2.

Table 2 Standards set in the County Leeds book

| Unit | Standard |
|---|---|
| 22 grains of wheat | = 1 Sterling |
| 20 pence | = 1 ounce |
| 12 ounces | = 1 pound |
| 8 pounds | = 1 gallon of wine |
| 8 gallons of wine | = 1 bushel |
| 8 bushels | = 1 quarter |

The Sterling then, was the basis around which the ancient English Troy Pound system was established (Hallock 1906: 32). The early Troy pound was nearly the same as the early German apothecary pound which was adopted almost exactly from 1/16 part of the Lesser Alexandrian Talent of silver (the German system was actually sixty-three grains lighter).

During the 13th century, England was rapidly becoming an active participant in interregional and international trade. Apart from language and cultural differences, interaction and competition with the Hanseatic League, Scotland, France, Flanders and Italy was additionally complicated because of the lack of any common metrological system. Because they were not bound by the statutes laid out in chapter Thirty-Five of England's Magna Carta, theses countries were free to conduct trade using a variety of different metrological standards. During this time however, England, Normandy, Flanders, Scandinavia and the Rhineland did have a common weight standard which was the mark, consisting of 216 grams. Although the coin values used in these countries differed, they could conduct commercial exchanges because they had a common ounce standard by which all other measures could be converted (Nightingale 1985: p.200). For England, international trade with European markets was the best means to achieve economic growth. This meant that a greater emphasis needed to be placed on establishing and enforcing British laws and standards. If she were to monopolize international trade, Britain had to first enforce laws regarding standardization at home. A document entitled the Composition of Yards and Perches was the first document intended to regulate measures of length and area as well as further define the use of the Sterling standard (Peterfreund 1989: 43). The Perch, as it was known, also reinforced the exclusionary intent expressed in the Magna Carta by prohibiting the use of all standards of measurement that were not in accord with the king's standard. Hereafter, linear measurement was to be based on a standard yard, which was three feet in length. Like the Roman system of measures, the foot was to be divided into twelve sections which were to known as inches. The effect of the Magna Carta and the Composition of Yards and Perches was twofold. First, the documents provided a badly needed framework in which the English system of standards could be developed. Naturally, English systems of weights and measures became the frames of reference and the metrological standards for interregional and international trade. Secondly, by establishing a concrete standard for measures, there was a movement to equate value with quantity and not quality. As Peterfreund observed, standardization was a necessary precondition to materialism and the mercantile frame of mind (Peterfreund 1989: 45-6). Hindsight allows us to see that these two documents shaped and focused the entire European frame of reference in terms of English standards.

By the late thirteenth or early fourteenth century, the Avoirdupois standard was extensively used in larger centers such as London, Winchester and Leeds. By the fifteenth century, English commerce utilized several weight systems including the Avoirdupois pound or libra mercatoria, which was based on the French sixteen ounce pound (Connor 1987: 197). While the Tower pound and Hanseatic League's system of weights were still used, the Avoirdupois standard became widely recognized as the single standard for commercial activities. The Troy pound was developed by Henry V for the Goldsmith's Guild of London for use in measuring precious metals and stones. The Troy standard is believed to have been adopted from the French commercial pound which originated at Troyes, France. Based on a sixteen ounce pound, the French also utilized the Livre de Troyes and Marc de Troyes systems which were compatible with the modern English Troy pound. In both cases, the English and French Troy standards were probably derived from the Roman weight of 1/125 of the Alexandrian Talent, or 5759.2 grains (Hallock 1906: 33).

Almost in defiance of the growing pressure for a single standard for weights and measures, shire towns and rich land barons struggled to maintain their private standards which effectively reduced the momentum of change in the direction of standardization. The English parliament passed a series of statutes in 1491, 1495 and 1496 which declared that all previous attempts at standardization had been ineffectual. The statutes also expressed the need for the eradication of all local customs in using and maintaining weights as well as regional variations. A new set of standards including weights, lengths and capacities was established for the Exchequer and stored at the Treasury in London. Copies of these new weights and measures were sent out to the forty-three main shire towns, and an appointed chief official was placed in charge of the task of upholding the new standard as well as the repair and replacement of old weights.

By 1590, England was well established as a major trade center. With rapid expansion in trade, most European countries maintained their own systems of weights for interregional use and adopted the Tudor Avoirdupois and Troy standards of metrology for international trade. These weights were viewed by many commercial trading houses as being the most reliable and universally acceptable units of measurement for mercantile activities. It is significant to note that in spite of her slow internal progress towards standardization in trade, other European countries viewed the English system as being the most accurate and efficiently administered system of weights and measures (Kula 198: 280). Internally, the problem of standardization of weights and measures was far from resolved. Major centers such as London, Exeter and Winchester were all utilizing variations of the Exchequer standards, which were the official standards for the realm.[2] Rich land barons and English nobility continued to maintain their own weight standards for regional trade. This had the effect of slowing down any progress towards standardization (Crellin and Scott 1969: 56). Tithes to the church and rents payable to land barons were often payable in weight systems that differed from the official standard. Weights were increasingly viewed as a "symbol of sovereignty" which municipalities and aristocracy alike guarded jealously (Kula 1986: 20). Equally covetous for power and revenues, guilds lobbied for the control and authority over the standardization of weight systems used in their trades. Revenues collected for the use of weights, and fines levied for their improper use, were substantial.

The Stuart accession to the throne in 1603 furthered the cause of standardization of weights by updating many statutes to abolish private use of weight systems and annulling old statutes which had been instituted to protect the interests of wealthy land owners. After 1603, there was a decline in the number of acceptable national standards, as towns, shires and land barons were forced to comply with the Exchequer standards. Over a period of three hundred years, the original intent of the Magna Carta was being realized as the Crown was increasingly able to force the acceptance of the Westminster standards, for both internal and international commerce.

---

[2] Differences from town to town were substantial. The Exchequer half-hundredweight contained 392,000 grains while the Norwich standard was 2679 grains heavier. The London and Exeter standards were approximately 1312 grains heavier than the Exchequer standard while the Winchester standard was 1313 grains lighter than the Exchequer standard (Zupko 1977, p.88).

While guild control over weight standards can be credited with eliminating many interregional differences, the ability of guilds to monitor and control their jurisdictions exacerbated all attempts at rapid conformity. The Goldsmith's Guild was granted monopolistic control over the Troy standard during the reign of Edgar III in 1327. They controlled the Troy standard at the Tower of London until 1679 and it was their responsibility to test coinage at the annual trial of the Pyx, which was conducted at Foster Lane Hall in London. At about the same time, the Pewterer's Guild was granted license to control the standards for pewterers and for weights made of brass. The Plumbers Company was granted exclusive jurisdiction over all lead weights in 1611, and it was their duty to "search, correct, reform, amend, assay and try all lead weight" in the city of London (Brewer 1853: 318). By the end of the first quarter of the seventeenth century, the Plumbers Company of London was also licensed to inspect weights made of iron. Because the Plumbers Company of London was a large, well organized guild, this may explain the sudden increase of stamped weights at this time (Crellin and Scott 1969: 53). Appointed representatives of the guild had the authority to enter any establishment and check all weights. If the weights were acceptable, it was the custom of the Plumbers Company to stamp each with the guild's "angel" seal. All weights that had been adjusted using plugs or weight adjusters were additionally stamped with a mark to distinguish them as having been corrected.

There has been some controversy over the origin of the use of the Archangel Michael figure which was used as the guild cipher of the Plumbers Company of London. The shop of Messr's De Graves and Company of London was founded in the late seventeenth century and remained a viable company using the angel figure until 1920 (Crellin and Scott 1969: 64 and Le Cheminant 1979: 289). The shop of De Grave and Company was located in St. Martin's le-Grand which was a walled precinct most noted for its large population of foreigners (aliens) and disreputable work force (Royal Historical Society 1961: 53).

An entry in the Brewers Records, dated at 1422, lists the mercers, grocers, drapers, fishmongers, goldsmiths, vintners, skinners, tailors, saddlers, ironmongers, girdlers, cordwainers, haberdashers, cutlers, armourers, weavers, fullers, dyers, plasterers and carpenters as being the twenty most important guilds in the City of London. In this list, which included 130 craft guilds, Pewterers, Plumbers, Joiners and Founders were amongst the top twenty five most important guilds within the city. Notably, the mercers, grocers, drapers and fishmonger were listed as being the most important fraternities, followed by the goldsmiths and plumbers (Unwin 1963: 370). Girdlers and cordwainers appear to have been the least influential guilds of the original Twelve Great Companies (Royal Historical Society 1961: 52). The London Guild system created many problems in maintaining craft standards amongst foreign artisans because it was difficult for them to be accepted as apprentices within the guild system which lasted up to seven years (Royal Historical Society 1961: 54). The Worshipful Company of Goldsmiths, which was in charge of monitoring the quality of all products made of gold and silver throughout the realm, noted that alien craftsmen came from France, Germany and Holland to work in London. Many of these aliens lived in St. Martin's le Grande and did not abide by the standards set by the guilds. Instead, they formed their own brotherhoods, which allowed them to maintain many of their traditional craft practices and styles alive (Royal Historical Society 1961: 59). Since there was such a high incidence of foreign craftsmen of all trades in this precinct, it is reasonable to assume that like goldsmiths, weightmakers also practiced their crafts outside of the standards and watchful eye of London guilds. The firm of Henry Neale (known as the Mathematical Practitioners of Tudor and Stuart England) and the firm of Young and Son produced the best quality boxed dispensing scales in the City. These firms were located in London's affluent business district of Bartholomew Lane and were close to the Plumbers Company Office at Westminster. Firms such as these located in the best business districts of London represented the core group of that city's artisans. It is possible that De Graves and Company may have brought the use of the Archangel Michael figure with them from an earlier weight making tradition in France. Because of the widespread use of the angel figure as a religious icon, its absolute origin remains unclear. There is no doubt, however, that it was the widespread use of this icon by the Plumbers Company of London that made it synonymous with the Plumbers Guild as well as a widely used symbol denoting quality.

The Founders Company was the last guild to be awarded metrological jurisdiction. From their Guild Hall offices in London, they controlled the standards for brass and bronze weights, taking this responsibility over from the Pewterer's guild. In 1614, they were given permission to assize and stamp all acceptable Avoirdupois weights used in the City of London as well as outlying areas within a three mile distance of the commercial center of the city. Weights that were acceptable to the Founders' standards were stamped with a ewer or laver insignia on their top surface. At the time of their appointment, the Founders Company was instructed to destroy all improper weights, and they were granted the authority to levy fines of twenty shillings for the possession of these improper weights. With the decline in popularity of the softer-bodied lead weights by 1700, however, the Founders Company became the most powerful guild in England and their jurisdiction over bronze weights was uncontested. It was not wise to question the authority of the guild's representative since any attempts at preventing him from carrying out his duties was punishable with a ten Shilling fine (Brewer 1853: 320). With the authority to stamp weights in both the Avoirdupois and Troy standards, their control over weights in seventeenth century England was absolute (Crellin and Scott 1969: 54).

By 1615 then, the three main systems of weight in active use were the Troy, Avoirdupois and Tower standards. In contravention of the parliamentary statutes from 1491 - 1496, several regional variations of the standard persisted in areas beyond the city limits, which in combination with the three official standards, kept English mercantile activities in a state of confusion. Major municipalities, free cities and noblemen, as mentioned earlier, frequently used weights that differed greatly from the Exchequer standard and many merchants maintained at least two differing systems - one that measured light for purchasing goods and a second that measured heavier for the selling of merchandise (Hallock 1906: 59). Attaining a single standard throughout the realm was no closer to being a reality in the third decade of the seventeenth century. This led to the implementation of the King's Beam system to ensure that a single standard for commerce was being employed. Lewes Roberts, a merchant in London, described the situation of standardization in 1638 stating that "the weights throughout Europe are all different, as well as from city to city and from mart to mart." He continued to note that in all major centers or "places of traffique there is found a weight authorized by the Magiftrate, which to alter or diminifh, is ever held a capitall crime" (Roberts 1638: 32-34). Other documentation confirms what Roberts had observed. The 1614 charter of King James I stated that:

> *"Whereas it is the part of our legal office to see our people be justly dealt withal and (among other things) to care that in weights and measures our subjects do deal one with another after a true proportion, wherein (as we are informed) there is a great abuse used, in that the weights made of brass are ill and unruly made, which groweth as well by the unskillfulness of the makers as by the corrupt and false making thereof" (Stock 1973: 39).*

Centralized weigh houses and stiff fines for land barons and landlords who maintained errant weight standards were the only effective means for conforming all of England to a unified set of weight standards. In 1670, Charles II directed that a set of brass weights of the Winchester standard was to be chained at every public marketplace and used as the sole standard for purchasing and selling. The Act of Common Council, dated at May 13, 1681 enacted laws which were designed to prevent unlawful and fraudulent sales transactions between London citizens and foreigners (Corporation of the City of London, 1698). With the implementation of resident brass weights at regulated weigh houses, foreigners found it increasingly difficult to conduct business, and many were forced to contract partnerships with freemen of the city (London). The Act of Common Council noted that these alliances were unlawful and inequitable to the freemen of the city. To prevent abuses of the system by foreigners, the city council ruled for the implementation of large fines against freemen who engaged in illegal partnerships with foreigners. After the first of January, 1699, all goods either sold or bartered by freemen and foreigners alike, had to be weighed at the King's Beam. Failure to do so meant that the seller was forced to pay a fine of thirteen Shillings and four pence for each five hundred weight or portion of that standard that was sold (Act of Common Council, 1698, & Atton & Holland 1967: 13). Although the Act was presented as a means to protect the freemen of London, the main concern was to collect fees, fines and taxes as a means to support the ever growing, poverty stricken population, who were forced to turn to poor houses. In a sense, the City of Freemen acted as Robin Hood, taking from the perceived rich in order to help the poor. The Act of Common Council also set some strict rates for duties to be paid for the use of the King's Beam which are listed in Table 3.

Along with the institution of the Common Beam, strict regulations were established for the care and testing of weights. The master of the beam was directed to examine the beam and weights against the standard at regular two-month intervals and a accurate log of all transactions was to be kept. Failure to record these transactions could result in either the levying of heavy fines against the weight master or his dismissal from this public post. In spite of these severe penalties, many infractions of the laws occurred.

But while penalties paid to the government were severe, guilds and fraternities also exacted heavy penalties from members who were in contravention of their rules. A German merchant was expelled from the Hansa trade network because he had paid the king's tax collector "more custom than was necessary." To compensate this unfortunate outcast, the Mayor of London made him a freeman of the city with all the rights and duties that status allowed (Atton & Holland 1967: 29). Ultimately, this was a favorable turn of events for the Hansa merchant since England and not the Hanseatic League was fast becoming the major trading force in Europe.

Table 3 Applicable charges for weighing.

| Weight | Charges |
|---|---|
| 28 lbs - 200 lbs | 2 pence |
| 200 lbs - 400 lbs | 3 pence per draught |
| 400 lbs - 600 lbs | 6 pence per draught |
| 600 lbs - 800 lbs | 10 pence per draught |
| 800 lbs - 10 cwt | 1 shilling, 4 pence / draught |
| 12 - 18 cwt | 2 shillings / draught |
| 18 - 22 cwt | 2 shillings, 4 pence / draught |
| 22 - 25 cwt | 3 shillings / draught |
| 1 bag of hops | 6 pence |
| every ton of iron | 2 shillings, 2 pence |
| every fodder of lead | 10 pence |

While there were three official weight standards which could be used for commerce, the Avoirdupois standard was most commonly used. Indeed, the largest denominations in the Troy standard were all but eliminated from the standard due to disuse, since the Avoirdupois standard was adopted for commercial and apothecary uses. Colleges of pharmacology taught that the Troy standard was to be used in the dispensing of all preparations, but the practice of dispensing of medications in the Avoirdupois standard was extremely common (Crellin and Scott 1969: 54).

With the strict ban of foreign standards in 1603, the Stuarts had set the stage for commercial standardization in which English weights and measures were to play the central role. During the next sixty years, many entrepreneurs were fined heavily for using weights and measures that deviated from the official weights of the Exchequer. All fines collected from foreigners and freemen alike were directed to the Treasury of the Corporation for the Poor, an agency of the City of London which financed and maintained poor houses throughout the city.

While English weights and measures were controlled and protected by guilds who enjoyed monopolistic controls over their jurisdictions, the manufacturing of weights in the Spanish New World remained under the control of the Crown, under the watchful eye of regional administrators. The relationship between the Sterling monetary unit and ounces unit of measurement was clearly defined in England although the association and control over the production of weights between English mints and weight makers is less clear. The Spanish Crown in contrast, maintained total control over the production of coins and weights. As early as 1537, the Spanish Crown demanded that all weights produced in Mexico were to be made of iron and each weight had to bear the seal of the city in which it was used. Between 1553 and 1559, the government mint at Mexico City manufactured weights, which bore the official mint stamp consisting of the letters "M." Variations of the "M" mint mark are illustrated in Figure 2. Often, decorative dots and circles were placed around the "M" mark. The third mark with the superscript "O" to the right of the "M" is probably an "O" inscription while the dot above the center "M" is probably an indicator mark showing which is the top of the letter.[3]

**Figure 2** Mexican mint marks:
Variations of the "M" mark.

Unlike England, the Spanish Crown maintained tight control over the production of coins and weights and guilds were never granted controls for the maintenance of weight systems. During this period, every city, town and *real de minas* or *mining center,* was to have a standardized system of weights and measures and the local government, by Viceregal authority, was to affix a seal of authenticity. But as in England, there was the potential for problems since each encomendero, alcaldia mayor and cabildo couldobtain the authority to use a private weight standard. It appears then, that with regards to her New World territories, the Spanish metrological system, like England's system, suffered from the same lack of standardization both regionally and internationally. It was only with a sweeping set of reforms during the rein of Philip II that a small degree of success in standardization was achieved in the Spanish New World (Kula 1986: 116).

**The German Monopoly of The Production Of Nested Weights**

In terms of construction and design, nested weights are one of the most complicated weight forms which was popular throughout Europe. For a period of nearly five hundred years between 1375 and 1850, England, France, Spain, Italy and Holland in part, all relied on the town of Nuremberg, Germany for the production of this important type of commercial weight. The height of production at Nuremberg was between the sixteenth and eighteenth centuries and master coppersmiths (Rotschmeidmeisters) in that city held the monopoly on the production of nested weights. While straight-sided, cast bronze nested weight sets with un-lathed sides were used in fifth-century Rome, it was the designs and craftsmanship of the German coppersmiths, combined with the need for a universally accepted metrological system in Europe, that brought Nuremberg nested weights to the forefront. Weight makers worked for several years and wrote extensive exams as part of their apprenticeship as journeymen. Indeed, many founders specialized in the production of weight sets of differing standards which were bound for foreign markets (Danforth 1954: 13).

Nested weights from Nuremberg were commonly produced as a group effort by artisans who specialized in various production techniques. As such, makers' marks found on some sets usually represent the name of a shop and not the name of an individual artisan. The process of making nested weights required a "former" or person pouring molten

[3] In his text entitled *The Coinage of the first Mint of the Americas at Mexico City, 1536-1572*, Robert Nesmith has included several variations of the "M" inscription on pages 45,51, 52, 55, 62 and 68.

bronze into molds as well as a "drechsler" who lathed the cast metal into its proper shape and proportions (Danforth 1954: 13). Apprentices stamped, polished and checked weight sets prior to shipment to foreign and domestic markets. Besides the numerous family shops which made nested weights, there were a large number of foundries operated by guilds, and apprentices in these shops were both numerous and important in the production of weights in these shops.

Crellin and Scott have observed that nested weights were the weight type most likely to be inaccurate when compared against legally issued standard. They suggest that in addition to being hard to adjust and correct, these weights were often not inspected and also subject to extensive use-wear. Henry Skinner, a manufacturer of weights and scales in London during the late 1860's, observed that Troy weight sets used by chemists, jewelers and pawn brokers were generally very light compared to the standard (Crellin & Scott 1969: 59 based on the Exchequer Standards Report of 1870: 21). In his opinion, a great proportion of these weights should not have been stamped as legal weights. Numerous weights in the Streeter Collection of weights as well as several in the Wellcome Collection are light according to the standards in effect when they were issued which suggests that quite possibly, use-wear and faulty production techniques are responsible for the large proportion of eighteenth and nineteenth century weights that are underweight when compared to the Westminster standards. Since the technology for making weights was passed on to successive generations of artisans through the guild system and weight sets themselves were used for extended periods of time, we can assume that a similar deviation from the official Troy standard probably existed in the late seventeenth century.[4] This trend of deviation from the standard may also be attributable to the fact that the Founders Company itself may have focused more on the regulating and stamping of weights and levying fines than on the actual production of weights, which were generally produced under contract by smaller shops. This is possible since the greatest potential for revenue was from the collection of fines and fees for stamping weights. Since several guilds, craftsmen, ironmongers, pewterers and plumbers all manufactured weights, there were numerous avenues along which deviation from the regulated standard could occur.

**Coin Weights**

Because of the close relationship between monetary and weight standards, there has been a close parallel between the insignia used on coins, commercial weights and coin weights. The first coin weights were produced for Henry III after 1250. Because of the soft nature of pure gold coins, bronze control weights were issued by the London mint as a means of monitoring changes in the coins from use-wear and clipping. During the reign of William III, demands for coin weights had increased to the point that private manufacturers were necessary to meet the demand for weights and scales. Until 1891, coin weights and scales were still considered necessary implements for day to day commerce in England. The Coinage Act of 1891 ended this reliance on personal money scales, however, because it stated that the English government guaranteed the face value of each new coin it produced, thus making it unnecessary to use coin weights and scales. To date, none of these coin weights have been identified from the artifact assemblage at Port Royal. Because of their prevalence in Europe, and mention in numerous Port Royal inventories, it would seem that we can expect to identify a variety of coin weights in future analysis. Since denominations of coins were in common use, coin weights would have served no practical purpose in day-to-day mercantile activities.

This brief overview of the history and development of weights has only touched on aspects of the developmental processes in England that have had an effect on the weight assemblage recovered from excavations at Port Royal, Jamaica. The greatest percentage of bronze and lead disc weights from the site bear the marks of English monarchs and the dagger insignia which was the emblem for the City of London. Because many of these weights were produced under guild controls, they also bear guild stamps of verification. The greatest proportion of weights recovered to date adhere to commercial standards as they were set by law during the mid seventeenth-century. While much of the British way of life was adopted in Jamaica, there was an increased need to legislate and strongly enforce laws for commerce in the colony. Because of her reputation as a wild and sinful colony, laws were enacted to guarantee settlers and entrepreneurs fair and equitable treatment while engaging in commerce in Jamaica.

---

[4] Skinner's testimony was used in the Third Report of the Commissioners Appointed to Enquire into the Condition of the Exchequer Standards: On the Abolition of Troy Weights, London, 1870, page 21.

**Law And Contemporaneous Society In Colonial Jamaica**

Many of the artifacts recovered from the 1692 provenance at Port Royal indicate that fashionable merchandise and goods were readily available in the colony. The contemporaneous nature of colonial Jamaican society is also evident in the laws under which mercantile activities and day-to-day life progressed. As discussed earlier, the Common Beam was well established in every major community by 1699 in England, complete with a maintenance structure and laws to ensure its proper maintenance and use. As early as 1681, legislation was in the planning stages to develop a similar system of controls over the use and control of weights in Jamaica. By 1700, a well-structured system was in place to control weights and measures on the island. Although many of the positions and titles outlined in legislation for the maintenance of weight standards differ from those used in England, the ideas and concepts for the management of weights are identical to the British system. Where masters of the Weigh House were solely responsible for maintaining weight systems and charging for transactions in England, the Town Clerk or "Clerk" assumed these responsibilities in Jamaica.[5] As in England, all transactions were to be logged in a ledger and strict fines were levied for infractions by either the Clerk or merchants. Similar to the English system, the party found guilty in any transaction at a scale was fined at the rate of one shilling three pence for each draught or measurement (Laws of Jamaica, Section CCXXXV 1681: 139).

The Town Clerk was also responsible for the "marking of every pot" or capacity container for which he was paid the sum of two pence for vessels ranging in size from a Gill (half pint) up to a gallon (Laws of Jamaica, Section CCXXXVI 1681: 139). Unlike the convoluted system of controls used in England in which guilds controlled the various weight systems, the Clerk in Jamaica was responsible for marking all weights, and he was paid at the rate of two pence for each verification stamp for weights ranging in size from one ounce up to the half-hundred weight. For each seal that he stamped on Yards or Ells as they were officially known, the Clerk was required to collect two pence.

The resident weight systems at wharfs, Stores and Houses (inns) were to be checked during the months of February and June by the Clerk of the Market. No fees were charged by the Clerk to check the weights and measures of "retailers, beings and inhabitants" since they were privately owned. Because they were privately owned, these instruments were only checked and if necessary, corrected once a year (Laws of Jamaica, section CCXXXIX 1681: 139). While no charges were paid for checking scales and weights, fees were paid for each transaction at wharfs, stores and inns.

Clearly, the brunt of Jamaican law was aimed at controlling commercial activities. Section CCXL of the Laws of Jamaica stated that no "Still-Yards" were to be used in any of the markets, retail stores or "Hucksters Shops".[6] Penalties of twenty shillings were levied for each offense of the misuse of a steelyard, which was a substantially higher fine than the twelve shilling fines, levied in England. Additionally, customers in Jamaica, had an added insensitive for reporting illegal trade. Ten of the twenty shillings paid as a fine were remitted to the Poor of the Parish while the remaining ten shillings were paid to the informer as a reward (Laws of Jamaica, Section CCXL 1681: 139). The remittance of funds for the maintenance and of poor houses was a direct copy of English laws which channeled all fines to the Corporation for the Poor, to maintain poor houses and feed the homeless and destitute (Roycroft 1698: 15).

In a special section entitled "An Act to encourage White Men to come to continue and settle in this Island," laws stipulated that all owners or people hired to operate public Wharfs, Barkadiers [7], Bays and Landing Places on the island

[5] During the seventeenth century, clerks were officers in charge of records and business transactions. The main criterion for clerks was the ability to read and write.

[6] During the seventeenth century, hucksters were retailers who operated out of small shops or stalls, selling small items of every description. Often hucksters were also called peddlers and hawkers.

[7] Barkadiers were the shallow slips provided for small sailing and rowing craft. Taken from the term barque or bark, Barkadiers were commonly used by fishermen, and coasting vessels, plying trade around the coasts of the island.

were to use an authorized set of weights and scales. The purpose of this "encouragement" was to guarantee that settlers and frequent visitors to the island were ensured fare and equitable treatment in their commercial activities (Laws of Jamaica, Section VII 1712). As the Act of Common Council had stipulated, the encouragement also stipulated that all transactions were to entered into a book and that a receipt was to be issued for each transaction. Failure on the part of the Clerk to provide these services was punishable by a fine of twenty shillings for each offense and dismissal, after repeated offenses.

**The Quagmire Of English Bureaucracy**

While the Troy standard was designed for use in both small and large commodity commercial transactions, popular usage in the markets of England reduced the Troy system to measurement of high value commodities such as gold, jewels and bread. The Avoirdupois standard became the standard for common commerce and the elaborate system of controls and verifications assigned to the guild structure was at best, only marginally successful in promoting standardization of weights and measures. As late as 1814, an inspector named John Warner observed that;

"there is a public office in Westminster, where they stamp all weights that are brought to them, whether just or not; and have been in this practice for years. I have now in my possession many weights of different sizes, some too heavy, others too light, all stamped at the Westminster Office, without their being put into scales" (Crellin & Scott 1969: 55).

Crellin and Scott also note that a large proportion of English bronze weights represented in the Wellcome Collection are grossly inaccurate, since many are underweight by at least twenty-five percent of their supposed standard weight (Crellin & Scott 1969: 55).

In his unpublished notes, Kisch observed that "of the three human inventions - counting, measuring and weighing - weighing is probably the youngest and the last to be integrated into the daily lives of people. This is probably due to the extent of involvement required by governments to supervise the activity." The quest for standardization of weights and measures is a relatively recent phenomenon which can be linked to expanded trade in Renaissance Europe. As noted at the beginning of this chapter, only the Anglo-Saxon world used the Troy and Avoirdupois standards simultaneously. Because these were two distinctly different weight standards, we must ask whether their use was the result of Britain's need for foreign markets and capital or because she was strategically located, serving as a cultural crossroads between Mediterranean and Northern European cultures. Undoubtedly, maintaining both metrological standards gave Britain a mercantile advantage in Europe and her geographic location exposed her to a culturally rich panorama of technologies and traditions from neighboring countries, which were incorporated as part of a developing and expanding English system.

Excavations at Port Royal have yielded an unusually high percentage of bronze disc weights. In a sense, this large representation of bronze weights indicates much about Port Royal. Because we know the time and date of the Port Royal earthquake, the high incidence of bronze weights confirms what Crellin and Scott had noted in that by 1700, lead weights were not as popular as they had been. This high incidence of bronze weights also points to the contemporaneousness of the seventeenth-century Port Royal community. Several bronze weights in the Port Royal collection deviate from the Westminster standards, both in mass and design. The Streeter Collection confirms what Henry Skinner the weight maker had observed, namely that bronze weights under the control of the Founders deviated greatly from the official standard during the seventeenth and eighteenth centuries.

Before discussing the relevance of these deviations and the archaeological provenance of the weight assemblage, it is necessary to discuss categorization of the weights as well as design features that were used in analysis of the collection.

## CHAPTER III

## CATEGORIZATION OF THE WEIGHT ASSEMBLAGE

The collection of weights from terrestrial and nautical excavations at Port Royal, Jamaica, is very impressive both in the number of weights collected as well as the diversity of the weight types. In total, 90 weights have been recovered from four main excavations including St. Peters Church, the New Street excavations, work conducted by Robert Marx (Marx 1967) and the ten years of nautical excavations conducted by Dr. D.L. Hamilton (Hamilton 1991) of Texas A&M University. While all the weights have been included as part of a comparative collection, only the weights recovered from the Texas A&M/INA (Institute of Nautical Archaeology) excavations are being used for spatial analysis since tight controls were maintained to record their provenance.

The quantitative breakdown of the weights is of special interest because, in many respects, the Port Royal site has yielded an unexpectedly high percentage of lead weights, as compared to other contemporaneous sites. For example, six lead disc weights from the sites account for 6.66% of the total assemblage of disc weights recovered from the site. Lead disc weights, as Richard Le Cheminant has observed, were never as plentiful or popular in use in England during the 1650's, although they are known to be contemporaneous with bronze disc weights (Le Cheminant 1979: 289). Because they were waning in popularity in the mid-seventeenth century, lead weights possibly found their way to the New World as cast-offs from closely regulated commercial activities in England. Another consideration is that since Port Royal was an extremely busy port supporting a large population of merchants and shop keepers, the use of these less popular weights was necessary since bronze disc weights, which were on the rise in popularity throughout Europe, were not readily available in quantity in the New World. While it was possible that some weights could have been cast in the colonies, lead and bronze metals were rare, necessitating that raw materials be imported to the New World for processing. A total of twenty-four bronze disc weights have been recovered, representing 26.66% of the total weight assemblage. A total of 44 weights, representing 44.44% of the total assemblage, have been recovered to date. This figure includes all lead bun, lead disc, dome and brick weights. A comparison of the percentages for all bronze weights and all categories of lead weights indicates a greater quantity of lead weights than has been suggested in the historic literature.

The last, and most probable explanation for this quantitative breakdown of the weights is that, as a maritime-based economy which was reliant on England as a trade partner, Port Royalists by stock and trade were forced to use government sanctioned weights. Portsmouth, Plymouth and the Thames River wharfs in London all supported similar successful maritime-based economies and in the seventeenth century, bronze weights were on the rise in popularity for commercial use in England. Because these towns were the ports of entry and export for England, government weigh houses and tax offices centered at these locations were amongst the first to receive standardized weights. It is probable that with the rise in popular use of bronze weights, which was the result of Charles II's decrees, there was an over abundance of accurate lead weights which found their way to the colonies. An investigation of the laws of Jamaica in a later chapter suggests that Jamaica, however, was possibly more strict in the regulation of weights and measures than her motherland. If there was an abundance of lead weights being sent to the colony, these weights were cast off because of their material composition and not because of their accuracy. Examination of Port Royal's lead weight assemblage indicates that this is true. The largest percentage of these weights are very close in mass to the official standard and many bear official stamps and ciphers to indicate their sanctioned use. It is interesting to note that by the end of the seventeenth century, England had largely done away with lead weights that adhered closely to the Avoirdupois standard in favor of bronze weights which tended to deviate greatly from the standard.

To facilitate an analysis of the Port Royal weight assemblage, the bun, disc and nested weights assemblage of the Streeter Collection of Weights, located at Yale University's Medical Historical Library served as an excellent comparative collection for analysis of the Port Royal collection. An investigation of Port Royal inventories between 1680 and 1710, as well as a brief investigation of the wills of known merchants and craftsmen was also conducted as a means of gathering data.

Over 140 entries from this thirty year period included scales and "stillyards" or steelyards. Appendix B, which is included at the end of this thesis illustrates the manner in which weights and scales were listed in inventories. Seventy-one entries indicate that the person owned "a pair of stillyards" and often a money scale. When a steelyard was mentioned, the inventory usually included a "parcel" or pile of weights. Very few entries are complete enough to indicate weight denominations and the number of weights listed in a deceased persons assets. Excerpts from the inventories of Charles Barre and William Davis are listed below. These examples serve to illustrate that it is impossible to determine types of weights and weight masses on the basis of the entries that were included in the inventories.

| | |
|---|---|
| Charles Barre | - one brass scale and pile of weights and a pair of stilliards. |
| William Davis | - one beame with scales and weights, two pair of small brafs scales and weights and two small pairs of scales and weights for money. |

It was no surprise to see weights and scales listed in the belongings of merchants and shop-keepers, but over eighty percent of the entries that included weights made no references to mercantile activities. Undoubtedly, even with the introduction of a community beam after 1700 AD, which served as the community standard, many households may have maintained weight sets and scales as their forefathers had done in Europe. Indeed, scales and weights were commonly passed from generation to generation as family heirlooms.

Weights are difficult to categorize because as is evident with the Port Royal assemblage, they are often used, adapted and altered from their original form. We can be certain that weights that had outlived their usefulness were salvaged for their lead content. Their secondary use value is another reason why lead weights might be desirable commodities in a maritime economy such as that of Port Royal.

To facilitate analysis, the weights assemblage was classified into eight subcategories, based on shape, materials and usage. These subcategories include (1) lead bun weights, (2) lead disc weights, (3) lead dome weights, (4) lead brick weights (trapezoidal in shape), (5) pail weights, (6) composite weights, (7) bronze disc weights and (8) bronze nested weights.

## Characteristics Of The Weight Types

### (1) Lead Bun Weights

Twenty-three bun weights, representing 25.55% of the total assemblage, have been recovered to date. Typically, these weights are round in shape and have either a flat irregular form (Figure 3) or a rounded dome surface as in Figure 4. Averaging 83.96 grams (2.96 pounds Avoirdupois) each, these weights have a less uniform appearance than lead disc weights which have a protective raised rim. Typically, few of the bun weights from the Port Royal site show signs that they bore the assize marks and stamps signifying that they were legal weights. Because they have no protective rim, most of these weights are badly worn and few of the verification stamps and ciphers that may have been issued with the weights have survived their heavy use-wear.

Many of the bun weights were crudely made by casting them in sand, which further explains their rough appearance. During the pouring process, slag and impurities in the metal floated to the surface, creating a thin layer of porous and irregular lead. As the lead continued to cool, the surface of the weight became slightly concave as well as crazed and cracked. After the molten lead had set, the weight was inverted so that its rough surface was on the bottom and the more uniform surface formed in the mold, served as the top of the weight. Almost all of the Port Royal bun weights have slightly concave bottom surfaces and traces of impurities suggest that most bun weights were cast using sand casting methods. Extensive use-wear, however, has obliterated some of these diagnostic features from the weights and in many cases, the bottom edges of many of the bun weights are severely disfigured.

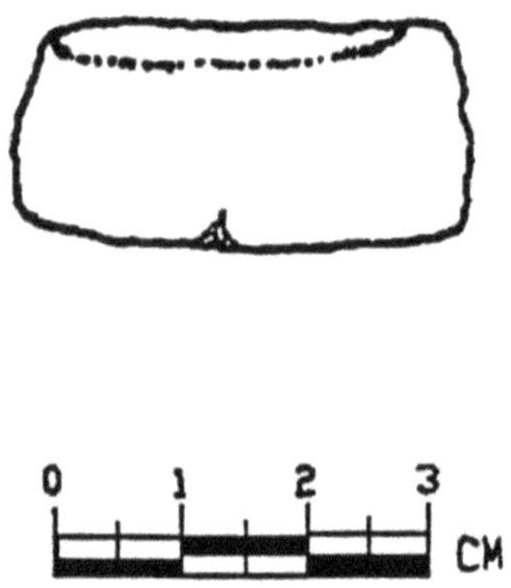


**Figure 3** A flat profiled bun weight with a rounded upper rim.

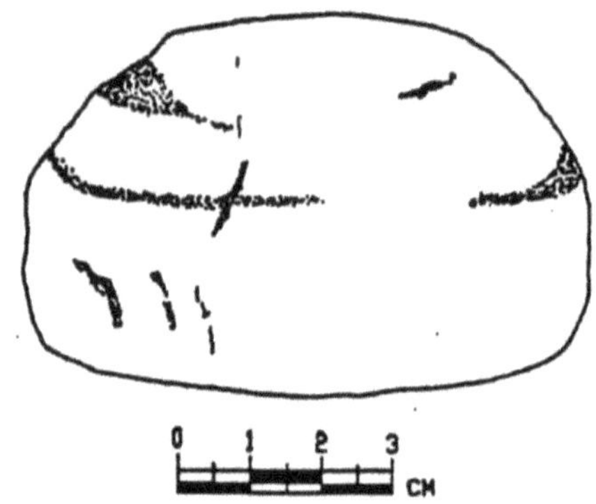


**Figure 4** A rounded or domed bun weight.

**(2) Lead Disc Weights**

Six lead disc weights from the site account for 6.66% of the total number of weights recovered. Unlike lead bun weights, disc weights have raised rims on their top surfaces. Only two of the Port Royal weights, PR85 1034-7 and PR85 1034-7 are well enough preserved on their bottom surfaces to indicate a possible second rim. Because of the softness of the lead, many disc weights show signs of heavy use-wear with rounded rims and distorted edges. In many cases, only portions of raised rims are visible. This makes their identification as disc weights more difficult because without their rim surfaces, they resemble lead bun weights. Similar to bun weights, disc weights are also manufactured by casting, using a press form to create a more uniform and well defined weight. As mentioned with bun weights, the process of pouring the lead causes a slight concave surface in which the edges of the weight appear to be slightly raised, once the lead has cooled. The raised edges of the weight, which are caused by the slight shrinkage of metal, can easily be mistaken as a rim on the bottom surface of the weight. The majority of lead disc weights have a single raised upper rim. Well preserved lead disc weights are more similar in design to bronze disc weights and many of them are stamped with assize marks and the seal of the Plumbers Company. Figures 5 and 6 show the top and bottom surfaces of artifact PR85 1034-7. On the top surface of this weight the rim is very visible and there is the faint outline of what is probably a circular assize mark. The bottom rim in Figure 6 is badly worn and a lead patch was attached in the recess created by the raised rim. This is a good example of one of the processes of weight adjustment used to bring an errant weight to a new standard which is discussed in chapter three.

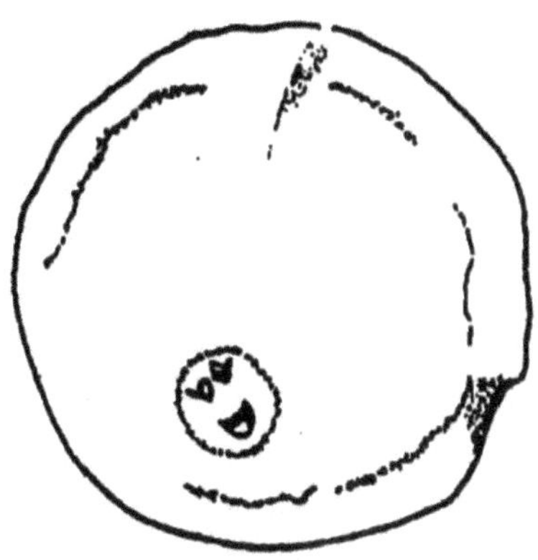

**Figure 5** Top surface of a disc weight with rim and cipher mark.

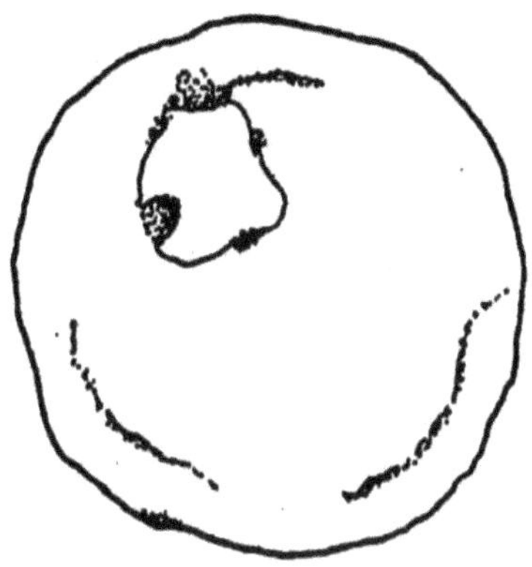

**Figure 6** Bottom surface with its weight adjuster.

**(3) Lead Dome Weights**

Like bronze disc weights, dome weights are diagnostically important for a number of reasons. While they are made of lead, and tend to lose their shape with extended use, dome weights are often recovered with legible assize marks positioned on their shoulders and other upper extremity surfaces. Because much of the top surface is protected by a large iron ring fastened to the body of the weight by a centrally located heavy staple, marks tend to survive well. The exception to this rule is that occasionally, stamps are placed in the path of the inclined handle of the weight, and over time, the weight of the handle disfigures any underlying ciphers. By the Ordinance of 1488, Henry VII decreed that every plumber in the community was to have a set of Exchequer-approved weights which were to be used as the standard for checking and correcting other weights. Included in their assemblage, each Plumber was to have a seven-pound weight which was to be used for the regulation of similar weights for commercial use. The Plumbers Company of London was granted jurisdiction over the production and keeping of lead weights and by their mandate, the guild monitored and stamped all acceptable weights as well as weights that had been corrected using plugs.

From the Port Royal site, seven dome weights have been recovered. With their iron ring handles attached, each of these weights would have weighed seven pounds Avoirdupois. These weights are PR85 1026-11, PR85 1026-11.1, PR85 1034-9, PR89 859-12, PR98, PR101. Weight PR85 1026-11 (Figure 7) is interesting in that on its side is the visible mark of melted lead, which appears to have been added to the weight to alter its mass. Because it would have been important to preserve the cipher marks which labeled the weight as a legal standard, it would have been necessary to adjust the mass of the weight by either adding weight to its side or its bottom.

PR 101, PR 1026-13 and PR89 859-12 all weigh close to 6.14 kilos or 13.50 pounds. With their ring handles and staples, these weights would have undoubtedly weighed fourteen pounds each. It is possible that weights cast as multiples of the seven pound standard were also viewed favorably according to King Henry's ordinance. One weight (PR98) measures 3.65 kilos or 9.24 pounds Avoirdupois minus its ring handle and staple. Calculating for the missing staple and ring handle, PR98 was undoubtedly a ten pound weight. All of these weights have indistinguishable stamps on their shoulders which may have been the marks of the Plumbers Company of London.

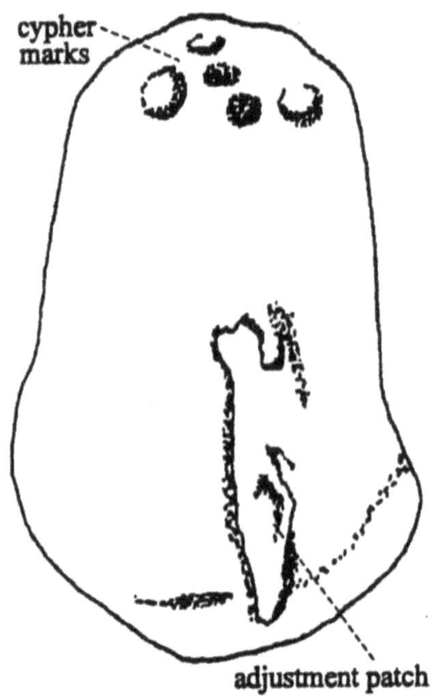


**Figure 7** PR85 1026-11 with cipher marks and adjustment patch.

**(4) Lead Brick Weights**

Like dome weights, brick weights were generally used on larger scales. These weights were transported and mounted on beams by means of a large iron ring handle. Like dome weights, ring handles on square weights were attached to the top surface by a large staple. While many sources have suggested that U-shaped staples were driven into the lead core of the weight to fasten the ring handle, examination of several Port Royal weights suggests that the ends of the staple were splayed or widened and then the handle and staple were positioned on top of the weight while the lead was molten. Using this handle, the weight was then either placed on a scale pan or suspended from a hook at the end of the scale arm or beam. Brick weights were commonly used in England up to the fifteenth century and while they continue in use to date, ornately decorated weights and a variety of new shapes were introduced in the sixteenth century (Crellin & Scott 1969: 59).

Brick weights are diagnostically important because marks and stamps, which are usually applied to the top surface of the weight abreast of the staple fastener, tend to survive well. Application of these stamps on the flat upper surfaces of the weight ensured that the mark could be struck more deeply into the body of the weight than could be achieved on the rounded surface of a dome weight. The squared design of these weights provided a larger surface areas on which stamps and ciphers could be placed. PR89 857-9 serves as a good example of the stamping practices on squared weights since this weight has three large "SP" stamp marks pressed into one of its side surfaces as well as guild marks and weight designation, stamped onto its top surface (Figure 8).

**Figure 8** PR89 857-9 side view showing "SP" markings.

Of the four brick weights recovered from Port Royal excavations, three have extensive markings which include weight designations, owners' marks and assize stamps. The largest weight, PR89 857-9 weighs forty-nine pounds after conservation, minus its mounting ring handle and additional weight adjusters. Allowing for these missing pieces, the mass of this weight would have been 55 pounds. Adjusters, in the form of iron shims, had been hammered into the top surface of the weight. From the inscribed Roman numerals "LV" on the top of the weight, we know that the intended mass of the weight was 55 pounds Avoirdupois. This weight also contains the impression marks of what are probably assize marks but unfortunately, none of these are legible.

Weights PR89 859-11(Figure 9) and PR89 859-13 (Figure 10) are similar in shape. The iron ring handle of PR89 859-13 was reconstructed by making a Hysol cast of the concretion that remained on the top of the weight. From the centrally located staple marks on the top of PR89-11, it is evident that this weight had a similar handle. This weight has a "VXX" inscription in its side surface which indicates that with staple and ring handle, the mass of this weight was intended to be twenty-five pounds Avoirdupois.

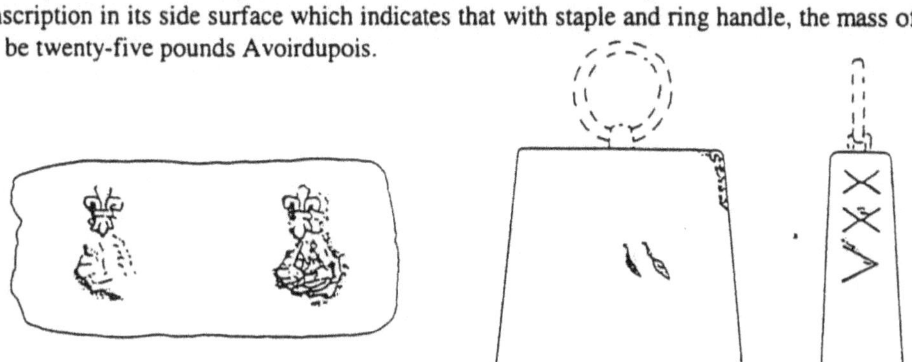


**Figure 9** PR89 859-11 weight with "VXX" inscribed on its side.

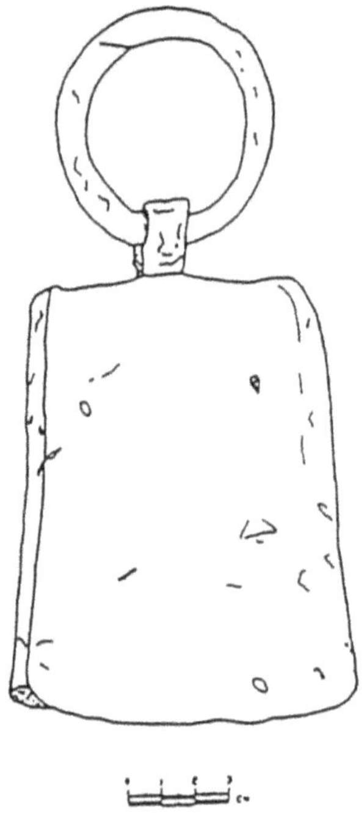

**Figure 11** Ship insignia on a coin weight (after Houben 1978: 8).

**Figure 12** Fleur-de-Lys and ship figure from PR89 859-11.

**Figure 10** PR89 859-13 with its ring handle

Both weights have distinctive markings on their top surfaces, located on either side of the staple attachment. These marks consist of a Fleur-de-Lys insignia surmounting a ship with rigging (Figures 11 & 12).

The first use of the Fleur-de-Lys and ships insignias on coin weights is dated at 1425. On these weights, either the ship or the Fleur-de-Lys was utilized, but the two were not combined as is evident on PR89 859-11 and PR89 859-13. The ship insignia on coin weights was always the dominant figure with a Fleur-de-Lys and a leopard included in miniature on each side of the mast (Houben 1978: 9). Since both weights adhere to the Avoirdupois standard, it is possible that the enlarging of the Fleur-de-Lys and the elimination of the leopard from the insignia was adopted in conjunction with the move to tougher standards by the Stuarts after 1603 and the granting of controls over lead weights to the Plumbers Company of London in 1611.

### (5) Pail Weights

Pail weights are a very unique category of weights. Serving many of the same functions as dome weights, pail weights are usually associated with the measurement of large mass quantities on commercial scales. Weight PR85 1026-13 (Figure 13) is such a pail weight and is distinguishable by its squared shoulder design, straight-sided body and large iron ring handle.

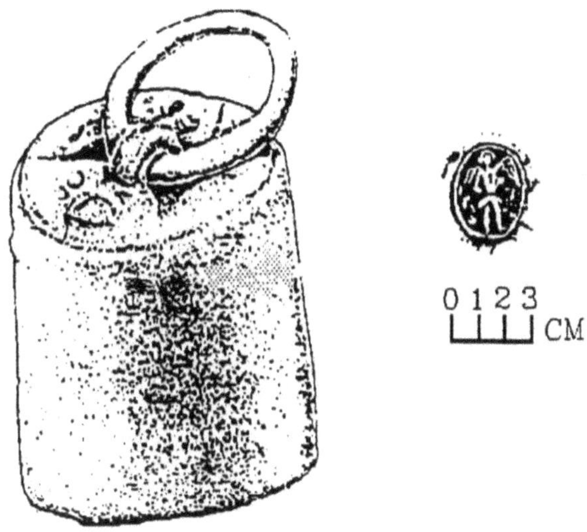


**Figure 13** PR85 1026-13 with its ring handle and Michael stamp.

To form the body of this weight, molten lead was poured into a cast iron pail-shaped container which when cooled, became the exterior surface of the weight. While the lead was molten, a ring handle and U-shaped staple with splayed ends were placed onto the open top surface of the weight and allowed to set in position. Once set, the handle and staple were firmly attached to the weight. Large iron handles fastened in this manner were common on dome and pail weights and were necessary for carrying the weight as well as suspending it from beams devices that did not use scale pans. To adjust the weight to standard, additional molten lead was poured onto the top surface of the weight as a means of trimming the weight to an exact standard. The distinct raised outline of poured lead on the top surface of PR85 1026-13 indicates that the weight was adjusted to standard in such a manor.

On the top surface of PR85 1026-13 are distinguishable marks including the Archangel Michael, which was the official stamp of the Plumbers Company, the dagger insignia of the City of London, and a Crown over "C" mark associated with the reigns of Charles I and II. As he often appears on weights, Michael is holding a pair of scales in his right hand, and often in his left hand, a sword. Several theories have been suggested for the origin of the Michael figure. The Archangel or "angel with scales," as he was commonly known, was possibly first used by the shop of Messrs. De Graves and Company, scale and weight makers in St. Martin's le-Grand. Thomas Overing of Bartholomew Lane (near the Royal Exchange) in London also used the angel figure although the Plumbers Company of London is most widely associated with its use (Le Cheminant1979: 290). Two variations of the Archangel Michael insignia are illustrated in Figure 14. The angel has been portrayed as both male and female, and while the figure is most commonly clothed in hose, it is occasionally dressed in a full pantaloons.

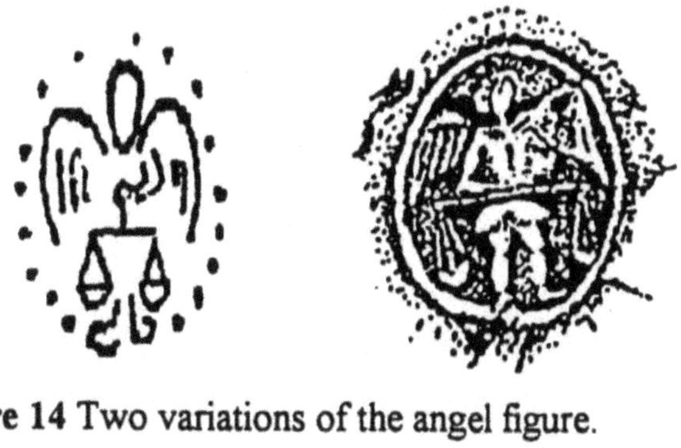

**Figure 14** Two variations of the angel figure.

**(6) Composite Weights**

Six composite weights, representing 6.66% of the total weight assemblage have been recovered from excavations at Port Royal. Five of these weights had cone shaped bodies and flat bottom surfaces and the remaining weight is round in shape with an iron ring handle.

To form the bodies of weights PR89 305 and PR83 423-40, sheets of copper were formed into cone shapes and then secured by crimping the overlapping edged into a long straight seam, running the length of the weight. An iron ring handle and staple fastener were then placed into a small hole at the tip end of the cone. Once the lead, which had been poured into the open wide end of the cone was allowed to set, the body of the weight was formed and the ring handle was firmly secured to the weight. Additional lead could then be poured into the base of the weight as needed, to adjust the cone weight to an exact standardized mass.

Instead of using lead to create an appropriate mass, cone weights PR87 426-6, PR 87 414-8 and PR 90 897-9.1 were filled with bound aggregate materials to form the core bodies of the weights. The exterior of PR90 897-9.1 was formed from a cone made of copper sheet. With its iron ring handle and staple in position at the tip end of the cone, a course plaster or sandy-mortar material was added to the cone, forming the core of the weight. The open cone edge of this weight has shallow straight sides at its base, and since no traces of metal were noted while conserving the body of the weight, it is possible that the weight had a wooden bottom, which was crimped into position to form a base (Figure 15).

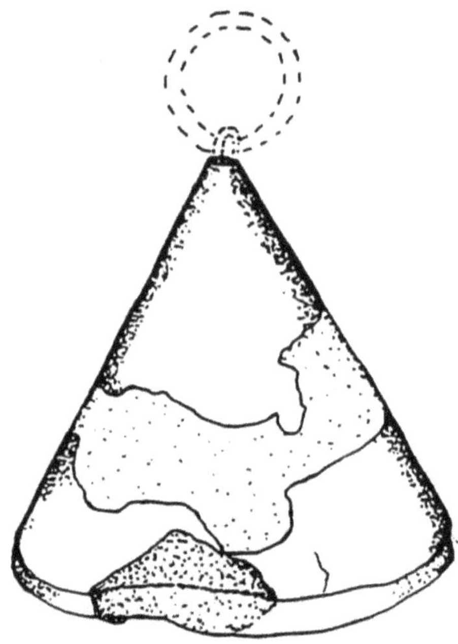

**Figure 15** PR90 897-9.1 composite cone weight.

Weight PR87 426-6 (Figure 16) was reconstructed from small fragments which remained of the original weight. Enough of the base survived that it was possible to determine that the circumference of the weight was almost identical to that of PR87 414-8. The angle of the sides of this weight were approximated from the small remnant of the cone surface that remained and the height of the weight was determined by using the angle of the body fragment to reconstruct the weight on paper. Although much of the core of the weight is badly deteriorated, remnants of aggregate material resembling course sand suggest that the core of the weight was similar to PR89 897 9.1. While the top of this weight was not recovered, it is most likely that this weight had an iron ring handle.

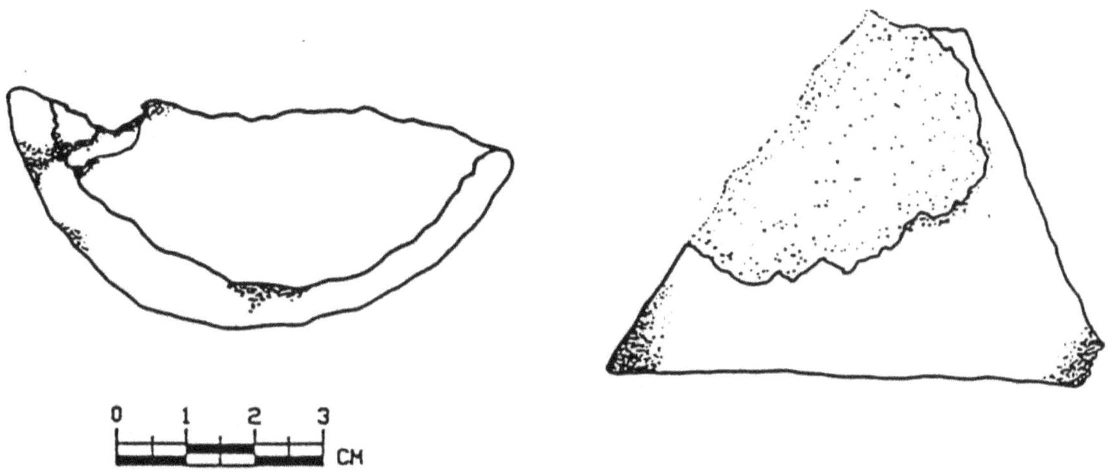


**Figure 16** Reconstruction of PR87 426-6 composite cone weight.

Weight PR87 414-8 (Figure 17) is one of the more unique composite weights. Because of its advanced state of deterioration, this weight was largely reconstructed using X-rays to record details from the small fragments remaining from the top of the weight. Discoloration in these concreted fragments suggest that the weight may have been formed from a copper cone, although none of the exterior surfaces of this weight have survived. Radiographs indicated that, instead of having an iron ring handle, this weight had a brass or copper wire handle, formed by knotting a length of wire and passing it through a small opening in the cone tip of the weight. Remnants of course sand or plaster in the upper tip of the weight suggest that the mass of this weight may have been formed using aggregate materials.

PR82 129.6 (Figure 18) is the most aesthetically pleasing of all composite weights in the assemblage. Formed as a ball counterpoise weight, the body of this weight was formed around a spherical hammered brass shell. An iron ring handle and splayed staple were placed in position covering one end of the ball and molten lead was poured into a hole in the opposite end of the brass casing, to form the core of the weight. Once set, the handle and staple were firmly affixed to the central core of the weight and any necessary adjustments to the mass of the weight could be made at the base to bring the weight to standard.

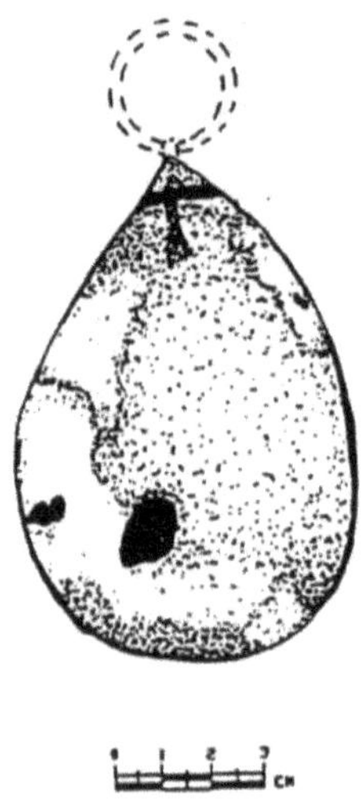

Figure 17 PR87 414-8 composite cone weight.

Figure 18 PR82 129.6 composite ball weight with iron ring handle and lead core.

All of the Port Royal composite weights were suitable for use as hanging counterpois weights (moveable weights) on small beams or as load weights when placed on a scale pan. PR90 897-9.1 was found in close association with a small scale arm (PR90 897 9.11) which was recovered as part of a large concretion in the hearth area of Building Five.

### (7) Bronze Disc Weights

In many respects, bronze disc weights are the most diagnostic of weight types in the assemblage because, due to the hardness of the metal and their ability to withstand extensive use-wear, both the weights and their markings are generally well preserved. The most common marks associated with bronze disc weights are monarch's marks in the form of an orb or crown over an initial, a dagger which signifies the City of London, an insignia stating the weight's standard as well as the ewer or laver mark of the Founders Company of London. Occasionally, owners marks are also included amongst the inscriptions on the upper flat surfaces of these weights. Figure 19 illustrates several of these marks.

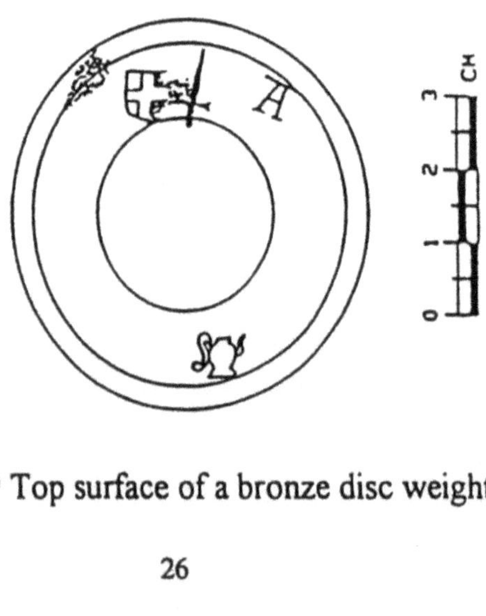


**Figure 19** Top surface of a bronze disc weight.

The "A" insignia, located in the upper right quadrant of the weight, denotes that this weight adheres to the Avoirdupois standard. The dagger and shield, which signifies the City of London, was also an adopted sign of many trades used to signify English quality. Many rural artisans allied themselves with London shops as a means of being able to use the insignia. The ewer stamp at the bottom of the disc is that of the Founders Company. PR156 as well as PR85 1024-7 and PR85 1024-7.5 possess unique markings because the ewer is pointing to the right, which is the reverse of the usual stamping process of the Founders Company. In a publication of the guild, the Worshipful Company of Founders illustrates three examples of the right-facing ewer, indicating that this particular version of the insignia was used from 1625 to 1685 during the reigns of Charles I and II. In 1685, the ewer was reversed back to its original left-facing position where it has remained ever since (The Founders Company 1991: 19). It is not known whether the reversal of the ewer was a miss-communication between guild offices or whether it was done as a means of observing the reigns of Charles I and II. This deviation from the standard usage of the guild insignia might also have been designed as a means of distinguishing which guild office had issued the weight. With increased pressure for the standardization of weights during the reign of Charles I, the demand for new weight increased, which placed greater pressures on guilds for increased production. The crest or shield included on the weight in Figure 19 was occasionally utilized instead of a monarch's initial.

**(8) Bronze Nested Weights**

Like bronze disc weights, nested weights recovered from excavations at Port Royal are diagnostically very important because their designs and inscriptions tell a great deal about the metrological and commercial systems from which they were derived. As noted earlier, Nuremberg was the major center for the production of nested weights between 1375 and 1850. While they did not control a total monopoly on the production of these weights, German craftsmen produced large quantities of weights of specified standards for foreign clients and inscribed them with appropriate stamps. The set of weights listed as being from the Institute of Jamaica are English in origin and are stamped with the Troy insignia. All cups in this set also bear Roman numerals on their top rims. Only four of these cups remain but because the outer cup and lid are in excellent shape, we can determine that the original set weighed two pounds Avoirdupois. The set consisted of a series of weights in 1, ½, 1/4, 1/8 and 1/16 pound cup sizes as well as a 1/16 disc weight to round off the set at two pounds.

Generally, the outer cup and cap of an English or German nested weight set is equal to the sum of the remaining cup weights. Since the sizing of the cups was usually based on fifty percent increments of the weight of the previous weight cup, determining the probable quantity and mass of the missing cups is simplified. Of all the nested weights recovered at Port Royal, the Institute of Jamaica nested weight set is the most massive in construction. With its plain decoration, this set represents the popular style used by English weight makers into the nineteenth century.

Several of the other weight cups have banding and decoration on their exterior surfaces. This style of finely lathed, ornate weight cups with pinned latches and hinges are more commonly associated with German styles of weight making. Notably, German weights usually have flat, roughly finished bottoms. These cups were turned or lathed with an external sprue which was eliminated during the weight adjustment phase of manufacturing when the bottom of the weight was filed to bring it to standard. Nuremberg weights from the seventeenth century seldom have kicks or turned indentations on their bottoms surfaces which is a characteristic trait of most English nested weights.

The lids of nested weight cup sets are diagnostically important because styles of construction and decoration vary between countries and regions. German weight cup lids are often very elaborately decorated with a handle mounted on top of a pair of pillars which were riveted to the lid. Additional decorations included cherubs, dolphins and elaborately carved lid hinges. Generally, English weight cup lids were considerably less ornate in their construction and decoration.

**Weight Adjusters**

Although they seem simple in their form, weight adjusters tell a great deal about the history of a weight. The weight of the adjuster plus its means of attachment to the errant weight, represents the mass required to bring a weight to standard. The percentage increase that the adjuster and nail represents can be used to determine the time frame for the correction of the errant weight, since new standards were well documented and are therefore, time specific. An example of this is the change in weight standard between the Florentine Libra (consisting of twelve ounces with an estimated weight of 339.84 grams) used for wool in thirteenth century England. There is only one-third of a gram difference between this old standard and the present Avoirdupois standard. Weight adjusters weighing one-third gram and of a style similar to those used in thirteenth century England most probably reflect this change in standard established by Edward I for use in the wool trade (after Connor 1987: 131-2).

Weight adjusters were commonly attached to host weights by means of a nail hammered through their centers, which anchored the adjuster to the main body of the weight. Examination of the weights associated with adjusters indicates that nailing or tacking additional mass to errant weights was the most popular means of adjusting weights in colonial Jamaica.

The process of making an adjuster was simple. By hammering a small piece of lead, its form was easily flattened and shaped to fit the contours of the host weight. The process of flattening the lead however, also produced the ragged-edged appearance, which was caused by stressing the metal into its new shape (Figure 20). Once flattened and shaped, the weight of the adjuster was easily adjusted by clipping its edges, similar to what was commonly done to coins.

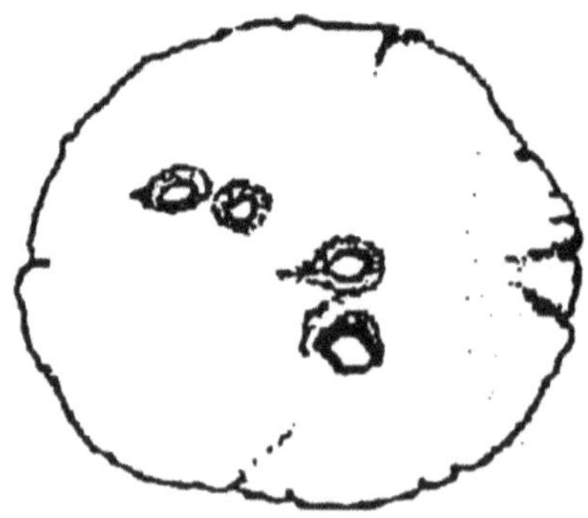

**Figure 20** Weight adjuster with cracked edges.

**Beams And Scales**

The most common type of measuring device in Port Royal was the common equal-arm scale or balance beam. Subsection CCXL of the Laws of Jamaica, dated at 1711, stipulated that "no Still-Yards be made use of in any Markets, Retailers or Hucksters Shops, under the Penalty of Twenty Shillings for each Time so made use of" (Laws of Jamaica 1711: 139). Still-yards or steelyards were unequal-armed measuring devices that were outlawed for use in English commercial transactions. The term yard is derived from the old English word gierd or gerd, meaning staff or measuring rod. In context of the laws as they applied to commercial activities, ells and yards were equal-armed scales and balances.

The term "still-yard" is the most commonly used term for weighing devices listed in inventories from Port Royal. While some inhabitants of the city possessed dessemers or unequal-armed steelyards which were illegal for commercial use, many of the still-yards listed were described as a "pair of still-yards" which conventionally, were equal-armed balance beams. The largest beams, used for commercial activities were commonly made of wood and iron. More precise scales used for pharmaceutical purposes and the measurement of precious materials were made of brass, steel, ivory and carved bamboo (Personal Inspection of the Streeter Collection).

As early Egyptians had discovered, the position and keen edge of the pivoting point and the attachment of the beam and its ability to move freely were critical in the making of scales or beams. Most seventeenth-century beams used a pointer arm which was attached to the free-moving balance arm of the scale as a means of indicating the horizontal position or point of equilibrium between the load weight and the mass of a commodity being measured. To ensure that the arm could move freely, a sharper fulcrum point or knife edge fulcrum was developed which increased the accuracy of scales dramatically. By the seventeenth century, this knife edged pivot point was positioned below the center line of the arm which was thought to make the beam more accurate. This is a popular design feature in use to date.

Steelyards are the unequal-armed beams which utilize an immovable fulcrum with a pan or hook attached at the end of the short arm for carrying a load. Suspended from the longer arm is a moveable counterpoise weight. When the beam of the steelyard is horizontal, the weight of the object being measured is indicated on an engraved scale at the point where the counterpoise is resting. The shapes of the counterpoise used on these beams are varied in form. Conical, cylindrical, zoomorphic and food-shaped counterpoise weights were commonly used. Many Roman counterpoise were formed in the representative busts of the heads and shoulders of military leaders and emperors.

Attachment of scale pans was also a serious consideration. Either three or four strings or chains were required to support each pan, depending on the size and weight of the materials that the scale was designed to accommodate. European designs incorporated four chains for heavy load measurement while sensitive scales, used for measuring precious smaller commodities, used three woven silk cords. A wide variety of materials have been used for the making of scale pans. Wood, reed baskets and mesh bags were among the most common materials used for early scales. European manufacturers used bronze, carved wood, silver, ivory, glass and porcelain. Regardless of the materials used, the most important aspect of pan construction was that both pans had to be of identical weight. The exception to this rule is seen in special scales that were produced with one specific function in mind. One pan of each set of Nuremberg ducket scales weighed exactly one ducket more than the other pan. In designing the scale this way, no additional weights were required for the measurement of these coins against the ducat standard.

To maintain the accuracy of a set of scales, it is important to relieve the tension that is applied to the fulcrum point of the scale when it is not in use. The easiest way to relieve pressure and maintain the fine edge of early scales was to either lay them down between usage or to suspend the beam above its resting point, thus relieving pressure from the fulcrum edge. The scales recovered from Port Royal all appear to be constructed using a simple "V-shaped" fulcrum point, made of steel. The large balance arm recovered during Marx's excavations is diagnostically important because it demonstrates some important construction techniques. After conservation, it was noted that the central fulcrum point mechanism as well as the attachments for pans and/or hooks at the ends of the arms were squared blocks of metal. During the manufacturing process, while the metal of the arms was red hot, these pieces would have been driven into their squared receptacles and then forged into a tight bond with the surrounding metal. Contraction of the metal of the beam arm while cooling would have ensured an extremely tight fit of all the components. Once in position, these squared blocks were honed and filed to finish the knife edge and hooks. While all of the beam arms recovered from the site show signs of extensive deterioration, many appear to have been manufactured using similar construction techniques.

No scales for weighing money have been recovered from the site to date, although they were specifically mentioned in several inventories. While Port Royal used denominated currency for day-to-day business, some areas of the colonial New World still relied on money scales for commercial transactions. One money weight, which was square in design, was recovered during early excavations at Port Royal. Unfortunately, only a sketch of the artifact remains and there are no notes regarding materials used in its construction or its weight. From Mayes excavations, a bronze disc weight with a large "M" initial stamped on each surface was recovered. This weight would have either been suspended by a small clip or placed on a balance pan, as a means of weighing coins.

Choosing categories for archaeological analysis is a challenging task because variation and specialization of weight types make it difficult to encompass all weight forms. All categories chosen for analysis of the Port Royal weight collection have been used for the analysis of other weight collections. There are very few special terms that differ from author to author, which made categorization easier. Where differences in terminology occur, I have included optional terminology in brackets.

# CHAPTER IV

## METHODS USED TO ADJUST AND FALSIFY WEIGHTS

Cipher marks indicate that the majority of the weights used in Port Royal were manufactured and assayed in England. In addition to guild ciphers, at least eleven weights show clear signs of having been adjusted or altered in mass, presumably to bring them to a new standard. Since the collection of lead weights all seem to adhere closely to the Avoirdupois standard, it is probable that other weights in the collection were also adjusted. Because of the softness of the lead and extensive use-wear however, some of the indicators of adjustment are not easily seen. Two important factors explain why lead weights were adjusted.

1. By 1615, guild controls and competition for the regulation of weights in the City of London was at an apex. The Exchequer, under pressure from the monarchy to accelerate the movement towards standardization, forced guilds to clamp down on the use of illegal and inaccurate weights. Tolls collected for the service of checking and stamping weights increased guild coffers substantially and fines collected for using unauthorized and substandard weights subsidized the growing number of poor houses within the city.

2. As a means to control trade and discourage abuses of the system by international traders, Charles II in the Acts of Common Council of 1681, set new laws and standards for regional and international trade. Apart from hefty fines levied by the guilds, all market places and weigh houses were to be furnished with a set of bronze weights and a measure made of brass which were to be chained to the building and were intended to serve as the model and standard for all citizens (Brewer 1853: 322). Legislation was also passed to ensure that all weights and measures were inspected twice a year and that town officials were responsible for destroying any weights and measures that were found to be defective. This had the effect of forcing guild representatives to compete with town officials over jurisdiction and the right to collect fines and fees for services. Competition also had an effect on mercantile practices in all communities since merchants were under the scrutiny of guild and city officials, and both vied for potential revenues. This made it increasingly difficult for merchants to use fraudulent weights in all major centers.

Listed in Table 4, are several weights from the site that were adjusted by using one of the techniques listed below:

1. nailing lead adjusters to the body of weights
2. adding melted lead to increase mass (PR85 1026-13)
3. adding large nails or spikes to increase mass
4. filing bronze weights to decrease their mass
5. lathing, etching and drilling bronze weights
6. clipping lead weights to decrease their mass.

Table 4 Adjusted Weights From Port Royal

| Identification | Method |
|---|---|
| Marx N.P. | - lead melted onto weight<br>- large nail added for mass |
| PR85 1026-8 | - edge clipped to reduce mass |
| PR85 1034-2 | - lead melted onto bottom |
| PR85 1034-6 | - lead adjuster nailed on bottom |
| PR85 1026-11 | - lead plug added |
| PR85 1026-13 | - lead added to top |
| PR 85 1034-5 | - weight adjuster nailed to top of dome |
| NS.2.A9.1b | - small hole drilled to reduce mass |
| PR90 964-7.1,7.2,7.3 | - each cup lathed to remove mass |
| PR89 897-7.1 | - cup filed on bottom to remove mass |
| Monte Christi Nested Weight Set | - bottoms filed to adjust to standard |
| PR85 1024-8.3 | - edge clipped to reduce mass |

**Means of Adding Mass to Weights**

The easiest means of adding mass to a weight was o simply hammer large nails into it until a desired mass was obtained. Once the new weight was achieved, the nails were usually clinched into the surface of the weight so that no metal protruded, forming a uniform surface on the weight. PR89 857-9 illustrates this method of weight adjustment quite well. There are a couple of disadvantages in adjusting weights in this manner. Over time, iron nails tend to rust, making the weight progressively lighter as the iron deteriorates. Continued deterioration of the iron causes the nail to become loose and eventually dislodge from the weight, resulting in an increased under-weight problem. This is problematic when analyzing weights since the size and mass of the nail or spike can only be approximated.

Another commonly practiced technique of weight adjustment was accomplished by attaching lead washers or weight adjusters to the body of the weight by means of a small iron nail or tack. This method had some advantages because the weight, nail and adjuster could all be placed on a scale and measured against an authorized weight. The lead washer would then be fine tuned in mass by clipping its edges before it was attached to the weight. The adjuster was then hammered onto the weight as a means of bringing it to standard. The soft lead adjuster was then form fitted to the weight by either lightly hammering it or through use-wear so that it followed the contours of the weight. Examination of PR85 1035-6 and PR85 1034-7 indicates that the soft thin metal of the weight adjusters have bonded with the malleable lead of the bodies of the weights. Over time, the iron nail fasteners have disintegrated but the adjusters have remained in their original fixed positions. Since the iron nails or tacks used to secure adjusters were very small compared to the spike or nail method of adjustment, the margin for error due to deterioration of the iron has been greatly reduced. A small squared hole on the top surface of PR85 1034-5 indicates that this lead weight either had a weight adjuster or had been adjusted using a small iron nail. There are no impression marks on the top surface of this artifact to indicate that an adjuster had been impressed into the fabric of the weight. Because the weight has a mass of exactly four pounds, it is probable that the weight had been adjusted using a small nail.

Plugs were commonly used to correct dome and square weights that were light in weight. Because of their softness, however, lead plugs were more difficult to use and not popular for weight adjustments. Iron plugs, or broad-headed stout nails, were more commonly used for adjustment and many included a cap or iron plate which increased the contact area of the plug with the lead weight. This ensured that the plug would remain in place and allowed for the addition of extra mass when it was required. The small lead weight illustrated in Figure 21 was recovered from the Saint Anne's Bay project on the North shore of Jamaica. Two large iron plugs with large plate bases were added to this weight to alter its mass. In both cases, the square dimensions of the plates are very large in comparison to the size of the plugs.

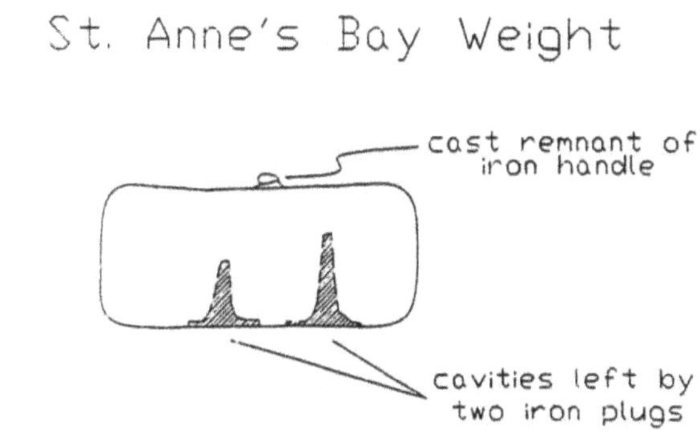


**Figure 21** Iron plugs with wide plate bases penetrate deeply into the body of the weight.

Artifact PR85 1026-11 was adjusted by adding a plug to the side of the weight. According to the charter of the Plumbers Company of London, weights adjusted or altered in any manner had to bear a stamp indicating that an adjustment had been made. Unfortunately, none of the four marks on the shoulder of the weight have survived.

Bonding melted lead to the surface of a weight was the least popular means of adjusting weights at Port Royal. The process of pouring molten lead onto a weight, presence of slag, spillage and any amount of lead left behind in the crucible made adjustments difficult by this method. Use-wear and the inherent softness of the lead make it nearly impossible to tell weather the bodies of the weights or the weight adjusters had been filed to remove excess mass. While it would be easier to remove excess lead with a file, it is doubtful weather this practice was common because as Joseph Moxon observed in "Mechanick Exercises, or, The Doctrine of Handy-Works," files were expensive, short-lived tools that were not made in the colonies, necessitating that they be imported from England (Moxon 1677: 68). Files were expensive and easily damaged by moisture and excessive humidity, necessitating that they be shipped to the colonies packed in beans or flour in the hope that they would remain dry. Files were too delicate and expensive to use for adjusting weights since lead shavings would render them useless quickly. Clipping or cutting lead would have been considerably easier and more effective for reducing the mass of an errant weight.

During the process of manufacturing and adjusting weights, excess metal was removed from overweight bronze disc weights by lathing or drilling the body of the weight to remove excess mass. NS.2.A9.1b has a single drill hole on its bottom surface that appears to have been made for just this purpose. This procedure was probably done at a later date than that of its manufacture because the hole was drilled inside of a nearly circular, etched groove, which was used for adjusting and fine tuning the mass of the weight during the manufacturing process. Unlike the process for adjusting lead weights, files were commonly used for adjusting bronze weights. The Monte Cristi nested weight set as well as numerous nested weights in the Streeter Collection, including bronze weights from the Early Roman Empire, exhibit clear signs of having been filed to adjust their mass. Weight cup PR89 897-7.1 from the Port Royal assemblage was filed on its bottom surface as a means of adjusting the mass of the cup. After conservation, the cup weighed almost 112 grams (3.95 ounces Avoirdupois) and undoubtedly, the original mass of this weight cup was intended to be 4 ounces Avoirdupois.

Although English and Jamaican legislation set strict laws for the use and control of weight standards, neither the Act of Common Council records for England or the Acts of Common Law for Jamaica stipulate acceptable procedures for the adjustment of weights. Such procedures appear to have been left to the discretion of artisans working in local shops. Blacksmiths, ironmongers, founders, pewterers and plumbers all practiced in the trade of adjusting weights in England (Crellin & Scott 1969: 62). There is no evidence to the contrary to indicate that craftsmen in Jamaica were not doing the same. Lead weights were produced in a wide variety of forms and styles and it was not until after 1700 with the increased authority of the Founders Company that there was a movement towards standardization of styles and forms of weights.

# CHAPTER V

# ARCHAEOLOGICAL DATA

In many respects the Port Royal Project is the ideal archaeological site since its location in Kingston Harbor, in close proximity to a functional lab and work area ensures that artifacts receive fresh water rinses and treatment without delay. The 1692 provenance of the site has been protected by layers of silt and staghorn coral that have prevented damage to the stratigraphy from natural forces such as storms, wave action and human intrusion. Because of a process known as liquefaction, house structures and surrounding roads sunk practically straight down, which had the effect of submerging large areas of the city, in a relatively undisturbed state. The Port Royal site is a catastrophic site (Hamilton 1991: 2) and unlike many archaeological sites whose artifact assemblages are representative of the refuse aspects of the people who lived there, the artifacts in many cases are whole and represent materials that were in day-to-day use, right up to the time of the city's destruction.

The same standards were applied to nautical excavations at the site as are used for terrestrial excavations. The work underwater however, is complicated by the necessity to use extensive machinery to supply air for breathing purposes as well as to operate dredge systems for clearing overburden from the 1692 provenance. Working in approximately ten feet of water, teams of student divers laid out ten foot grids constructed of ABS plastic and excavated the foundations of several structures. Throughout the process of excavation, extensive field notes on the provenance of artifacts and architectural features were recorded to facilitate the job of reconstructing the site on paper. Dr. Hamilton conducted nautical excavations at the Port Royal site between 1981 and 1990. From 1990 to the present, research and work has continued on the backlog of artifacts and information which have yet to be processed. Given the remarkably large size of the artifact assemblage recovered from the site, the job of conservation and analysis will take many more years.

All of the weights recovered during TAMU/INA excavations come from four separate areas within the excavated structures. These areas are:

1. Building One

2. Building Three

3. The Ship (one weight)

4. Building Five including the yard of Building Seven

Figure 22 (over leaf) is a site plan of TAMU / INA excavations at Port Royal. Although only one weight was recovered in direct association with the ship, its beveled edges and squared form are unique as compared with other weight forms associated with the site.

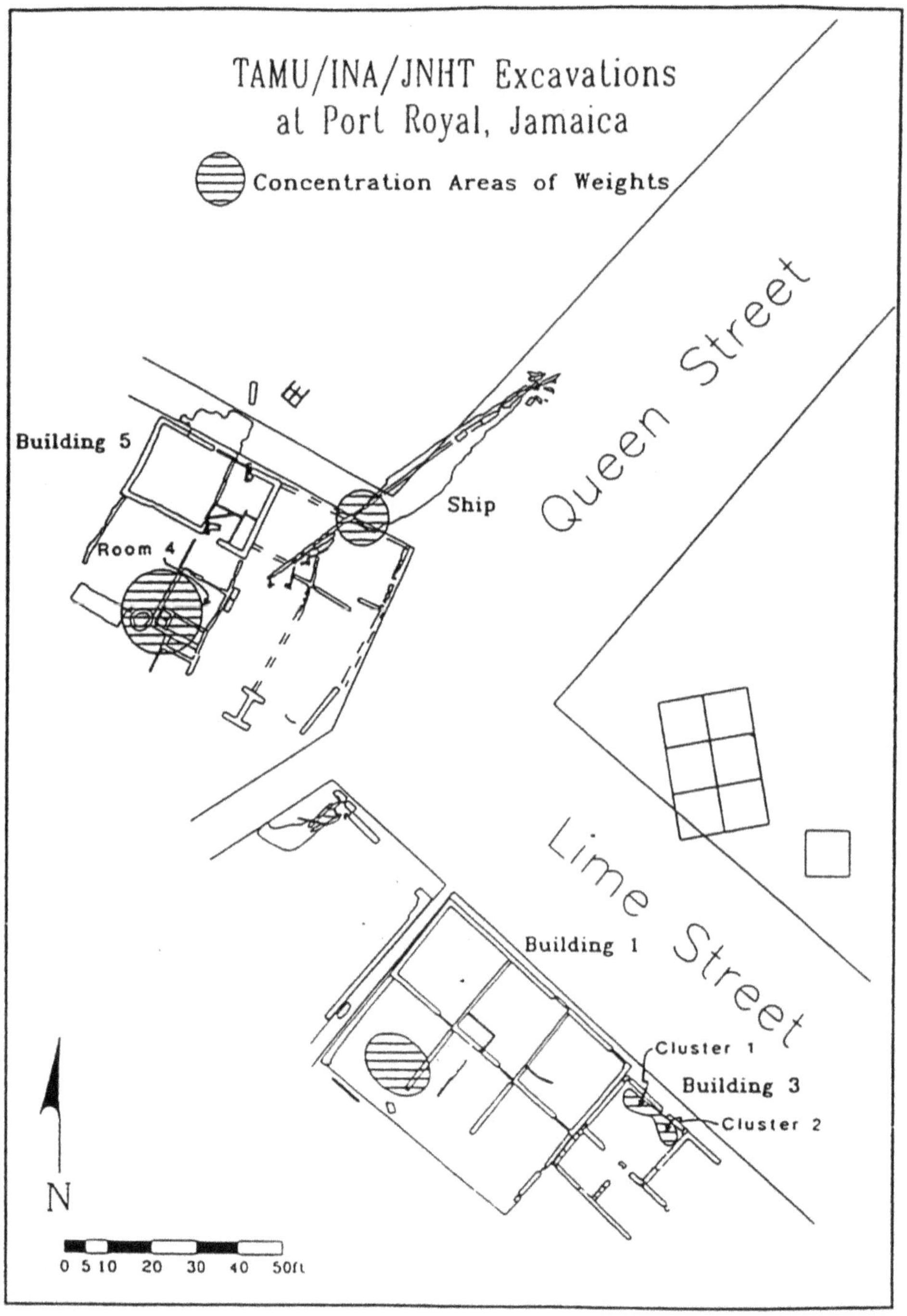


**Figure 22** Concentration areas of weights, Port Royal site plan.

In total, 90 weights have been recovered from Port Royal excavations. A breakdown of the weights shows that the largest proportion of weights were made of lead totaling 40 in number, not including weight adjusters. This represents 44.37% of the assemblage. Forty-two bronze weights, representing 46.66% of the total weights were also recovered. Six composite weight, representing 6.66% of the assemblage and two pail weights were also recovered. Pail weights represénted 2.22% of the total assemblage. Data are included in Figure 23.

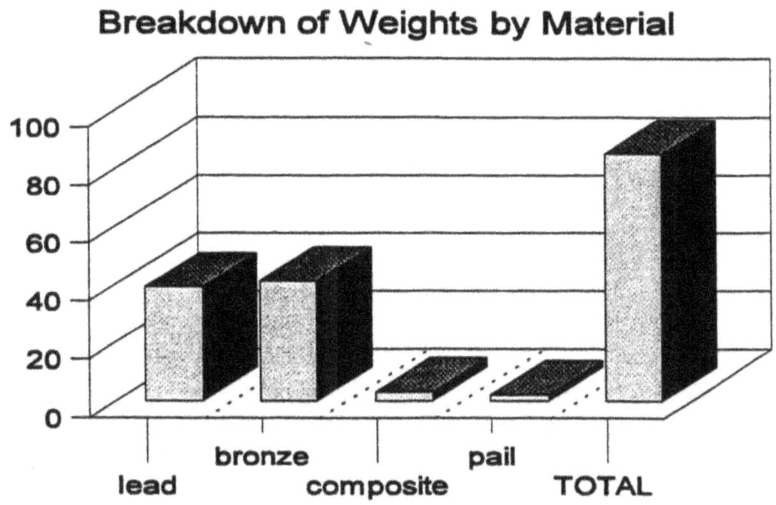


**Figure 23** Percentage Breakdown of Weights By Materials

Lead bun, lead disc and bronze disc and nested weights are the most common weight types in the Port Royal weight collection. Broken down by major weight categories, twenty-three lead bun weights (25.55%), 6 lead disc weights (6.66%), 7 lead dome weights (7.77%), 4 lead brick weights (4.44%), 2 pail weights (2.22%), 6 composite weights (6.66%), 24 bronze disc weights (26.66%) and 18 bronze nested weights (20.00%) have been recovered.

**Weights from Building One**

Building One is comprised of six rooms. Rooms 1, 3 and 5 were the original building while rooms 2,4 and 6 were additions. Room 1 appears to have been used as a cordwainer-shoemaker's shop. In this room, numerous scraps of leather and tools appropriate for making shoes were recovered. Because of the high concentration of bones found in Room 2, this room may have been part of a butcher's shop. In rooms 3 and 4, remnants of furniture were found and in rooms 5 and 6, remnants of a large quantity of un-smoked white clay pipes were recovered. Rooms 5 and 6 also had a large number of whole onion bottles. These artifacts may indicate that these two rooms were a vintner's or tobacconist's shop.

A total of three weights and two scale arms were recovered from Building One. Scale arm (PR85 1022-5), counterpois weight (PR82 129.6) and a lead bun weight (PR86-144-6) were recovered from Room 2 of the building and a larger scale arm (PR85 1025-7) and a composite weight (PR83 423-40) were found in Room 4. The locations of these artifacts are illustrated in Figure 24.

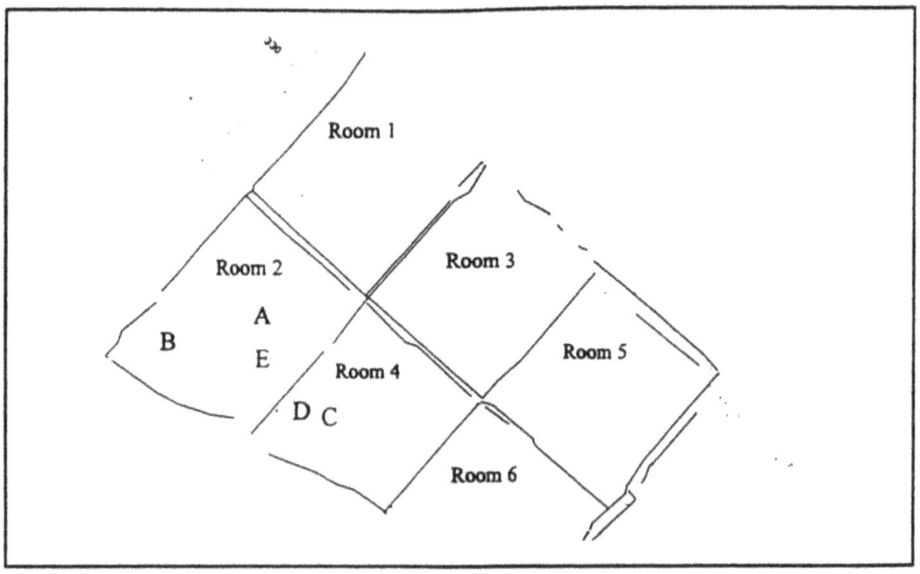


**Figure 24** Location of weights in Building One.

**Key to Figure 24**

| Description | Weight Type |
|---|---|
| A | scale weight - PR82 129-6 |
| B | scale arm - PR85 1022-5 |
| C | scale arm - PR85 1025-7 |
| D | composite weight - PR83 423.40 |
| E | bun weight - PR86 144-6 |

All weights and scale arms recovered from Building One were common tools used for commercial activities. PR82 129-6, is a ball weight of 2.24 kilograms mass (4.94 pounds Avoirdupois), minus its mounting ring and staple. The weight consists of a spherical brass casing with a cast iron staple and ring positioned on its top. Lead was poured into the weight and when cooled, the splayed edges of the staple firmly fastened the iron ring handle to the lead core. Because the bottom of the weight was open, fine adjustments to the mass of the weight could be made by either adding or eliminating lead from its base. The iron mounting ring was reconstructed using epoxy resin to cast the cavity left in the concretion from the corrosion process. These features are visible in Figure 25.

**Figure 25** Ball weight, Pr82 129-6.

Including the ring, this was most probably a five pound weight used on steelyards or suspended by a hook from a balance scale. The base of this weight has a slightly flattened small area of lead and with an iron ring handle, the weight would probably be very unstable when placed on a scale pan. A weight of this size was too large to have been used on money scales. The remaining weight from Room 2 of Building One is a small bun weight of .21 kilos (.47 lbs.) that was probably used on a small scale or balance beam. The only distinctive features of PR86 144-6 are a series of deeply incised lines on its bottom surface. All surfaces of this weight are extensively pitted, suggesting that the weight was constructed by pouring poor quality lead into a sand mold.

Composite weight PR83 423-40 was recovered from Room 4 in Building One. The construction of this weight and PR305 (Marx N.P.) are almost identical. The copper cone which houses the lead core of both weights was created by folding a sheet of copper into a cone shape and then folding the edges of the sheet to create a single seam running the length of the body of the weight. In both cases, a splayed staple fastener and iron ring handle were positioned through a hole in the pointed tip of the cone and lead was poured into the open end of the cone to create the core of the weights. Because of the softness of lead, the copper sheathing was a functional means of preventing wear to the weights as well as making them more aesthetically pleasing in appearance. Once cooled, adjustments to the weight could be made by either adding of clipping lead from the core to bring the weight to a standardized mass.

All three commercial weights appear to have adhered closely to the Avoirdupois standard, although in the case of the composite cone weight and ball weight, this is speculation based on the probable weight of the iron fittings which have deteriorated. Since the system of inspection for weights has been described as totally inadequate and numerous weights at the Westminster office are known to have been stamped without ever being checked, a wide range of variance from the Exchequer Avoirdupois standards can be expected. Crellin and Scott have noted that seventeenth-century bronze weights tend to be overweight by as much as twenty-five percent of their target weight (Crellin & Scott 1969: 55). Lead weights are usually much closer to the Avoirdupois standard. This overweight problem may not have been viewed as a great problem amongst local merchants, however, since many of the scales in use were not sensitive enough to measure small variations from the standard. A scale calibrated to accommodate these heavy weights would probably accommodate a legal weight within an acceptable range of variance. PR86 144-6 and PR82 129-6 are only slightly underweight according to the Avoirdupois standard and most assuredly, both weights were five pound weights. The weight of composite cone weight PR83 423-40 is more difficult to determine. With staple and iron ring handle, however, this weight probably had a one pound mass.

### The Large Collection of Weights from Building Three

The excavation area of Building Three consists of one entire rooms and two partial rooms. Weights in Room 1 of the building were concentrated into two distinct areas along the North interior wall of the structure. Two scale arms and a single scale pan were also found in associated with these weights. In this same room, one of the largest concentrations of un-smoked white clay pipes and numerous whole onion bottles were also excavated. Considering the nature of the artifacts recovered, Building Three may have been used as a vintner's shop or as a storage and preparation area. Locations of Cluster 1 and Cluster 2 are illustrated in Figure 26.

Twenty-one lead weights, 9 bronze weights, one pail weight, 2 iron scale arms and one scale pan were recovered from this area. Interestingly, the lead weights adhered closer to the Avoirdupois standard with fifteen of nineteen weights being within one percent of the standard weight. Two of the lead weights, PR85 1034-6 and PR85 1035-7, showed signs of having been adjusted to bring them to standard. PR85 1034-6 is a well preserved disc weight with a lead adjuster nailed to its bottom surface. PR85 1034-7 was adjusted in weight by melting a quantity of lead unto its flat surface as a means of adding mass. Because the surface of this weight is rough in texture, it is possible that a small nail was not necessary to fasten the weight adjuster to the weight.

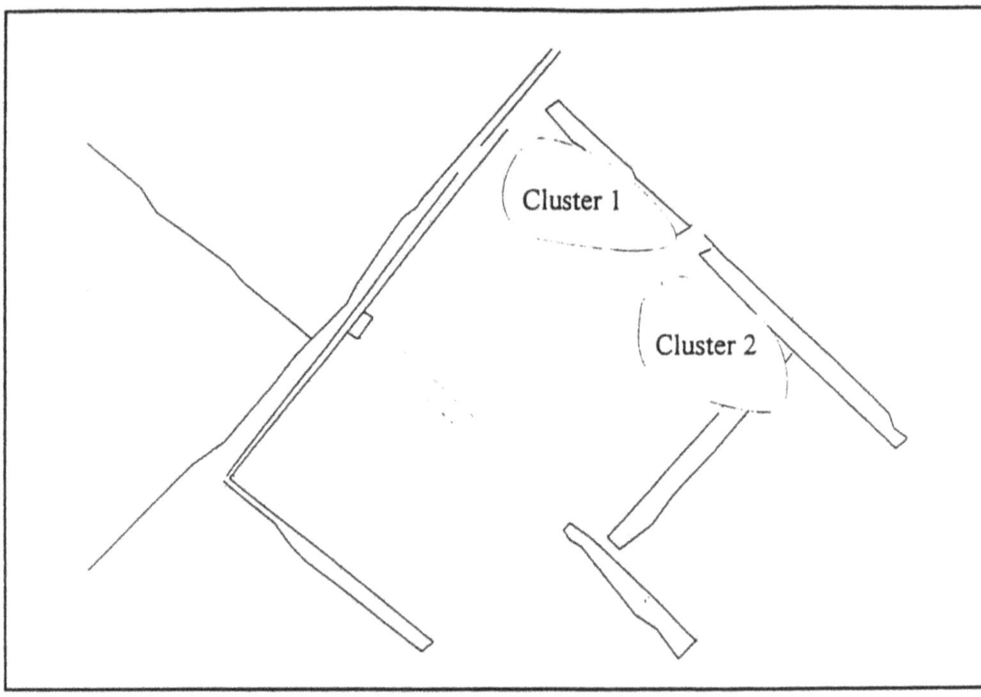


**Figure 26** Location of weights in Building Three.

Eight of the weights recovered were weight adjusters. All of these weights have a single squared hole running through their centers and some were found in context nailed onto the surfaces of other weights. One adjuster in particular, PR85 1024-8.3, has had its edge clipped as a means of altering its mass. In order of increasing size, the weights of these adjusters are .572, .578, .594, .655, .745 and .864 ounces. At first glance, these adjusters may seem to be a unique set of weights. Three of these adjusters are in approximate 1/10 ounce increments while the other three are in 1/100 ounce increments. The best explanation for the apparent incremental masses of these adjusters is that they may have belonged to an actual set of weight that had been adjusted. This would necessitate the addition of lead adjusters that, like the individual weights in the set, were incremental in size.

Four of the nine bronze disc weights in Building Three adhered to the Avoirdupois standard and included the "A" cipher mark on their top surfaces. PR85 1024-7, PR85 1024-7.5, PR 85 1026-9 and PR85 1034-5 also include the dagger insignia which signifies London as their city of origin. All of these weights are slightly under weight by the Avoirdupois standard. PR85 1026-7 is underweight of the three-quarter pound weight standard and has none of the necessary ciphers to indicate that it was a sanctioned weight. PR85 1034-6 does not have an Avoirdupois mark but does include two daggers and an orb. Currently weighing 7.3 ounces after conservation, this was most likely an eight ounce weigh. While a weight adjuster is present on the weight, the nail fastener is no longer present. Disc weights were commonly made in ½, 1, 2, 4, 5 and 7 pound increments of size and were considered to be the standard for the measurement of goods sold by merchants. Le Cheminant observed that these weights had been commonly found around the Thames River wharves, as well as in many market places (Le Cheminant 1979: 29).

The evolution of the coat of arms for the City of London involved several transitional phases. To date, there is no complete record of the transitions that the cipher went through from its initial use. In his article entitled "Interesting Marks On Weights," Graham notes that the dagger was used as the royal cipher for Avoirdupois weights at London's Guildhall until 1826 and after that date, the dagger was included in a full coat of arms (Graham 1979: 24). One of the bronze disc weights recovered during Marx's excavations (PR156) includes the coat of arms in conjunction with a large dagger. This weight possibly represents a transitional design leading to the later coat of arms for the City of London. Interestingly, the ewer on this weight is facing to the right, which sets its date of manufacture between 1625 and 1685 (The Founders Company 1991: 19). Many artisans from areas outside the jurisdiction of London's guilds maintained alliances with London craft houses and merchants and many of these used variations of the dagger insignia which became synonymous as a symbol of quality, as well as the mark of the city of London.

Dome weight PR85 1026-11.1 is an interesting weight recovered along the wall in Building Three. Apart from a centrally located hole which housed an iron handle and staple fastener, there are four additional indentations containing faint markings on the shoulder of the weight. The side surface of this weight, visible in Figure 27, has been altered by melting additional lead into a slightly raised clump on the side of the weight. With staple and Iron ring handle, the mass of this weight was most likely 7 pounds Avoirdupois.

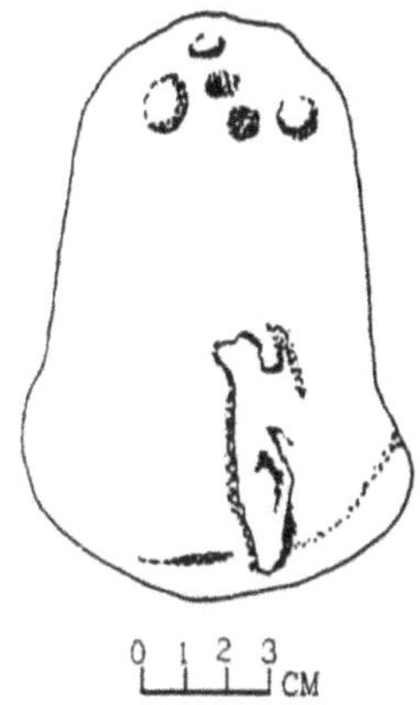


**Figure 27** Pr85 1026-11 with cipher indentations on its top surface.

One of the partially visible marks on the top surface of this weight appears to be an adjustment mark used by the Plumbers Company to indicate that the weight had been adjusted to a new standard. It was necessary for the guild to mark weights that had been visibly altered accordingly, to ensure their acceptance for use. Three of these seven pound weights were recovered at Port Royal and they are significant because by the Ordinance of 1488, Henry VII had commanded that every plumber in the community was to have a set of Exchequer - approved weights, including a seven pound weight.

Pail weight PR 85 1026-13 is an diagnostically important weight. The top surface of this weight contained large iron handles as well as several ciphers used by the Plumbers Company. Included on this weight is the mark of the Archangel Michael, a dagger representing the City of London and a crown over "C" mark of King Charles.

Pail or dome weights were extensively used in the early days of the Plumbers Company of London. These weights always included the angel figure with a pair of scales in his left hand and the Guildhall dagger or a sword in his right hand. Although they had been popular circa 1615, these weights were banned from use in 1835 because of their propensity to rust and disintegrate (Graham 1979: 11). As their iron jackets deteriorated, pail weights tended to slowly drift away from their established standard weight. Two iron scale beam arms were also recovered from Room One. PR85 1025-7 is the larger of the two beams and would have been capable of weighing heavier loads than PR85 1022-5. While the beam is badly deteriorated, it was possible to determine that the beam arm was forged from a single piece of metal. PR85 1022-5 is smaller in width and thickness and enough remains of the knife edge to indicate that this part of the beam had been made from squared stock which was pressed into a squared receptacle in the beam and forged to create a tight fit.

Analysis of the breakdown of weight types indicates that lead weights were in predominant use for commercial activities in Building Three. As Figure 28 indicates, only lead weights show signs of having been adjusted to a different standard as is evidenced by the fact that several have had weight adjusters and plugs added to increase their mass. Not including the two balance beams and weight adjusters, all of the weights appear to have adhered to the Avoirdupois standard, in both markings and mass. Lead bun weight PR85 1026-8 appears to have been drastically altered by clipping a large portion of one edge, reducing its mass to 1.75 pounds. Because this weight has the remains of two cipher marks and two incised slash marks on its top surface, it is possible that this was a two pound Avoirdupois weight that had been either altered for a specific purpose or used for some other non-mercantile activities. Avoirdupois standard weights from Building Three account for 80.95% of the total assemblage and are contained almost entirely in room one.

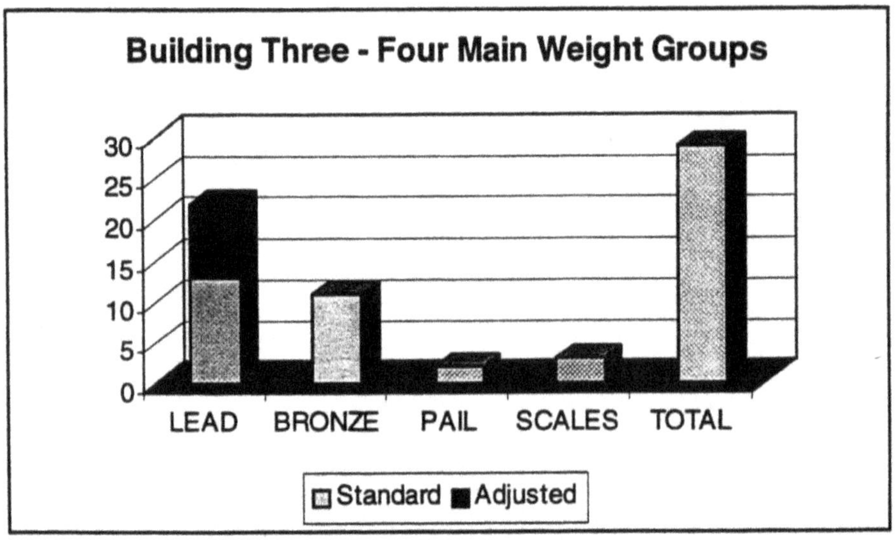


**The Ship and its Single Weight**

Only one weight was found in association with the ship. PR90 843-5 was retrieved between two frames of the vessel and its position is illustrated as letter "X" in Figure 29. The bottom surface of the weight contains pockets of slag metal, formed by impurities in the lead floating to the surface of the cast, while it was cooling. The most important feature of this weight is its beveled upper edge, which according to Crellin and Scott, is a rare design feature for lead weights of this time period (Crellin & Scott 1969: 59). PR90 843-5 weighs slightly more than 2 pounds 6 ounces Avoirdupois and was a common weight form used for measuring wheat for baking bread.

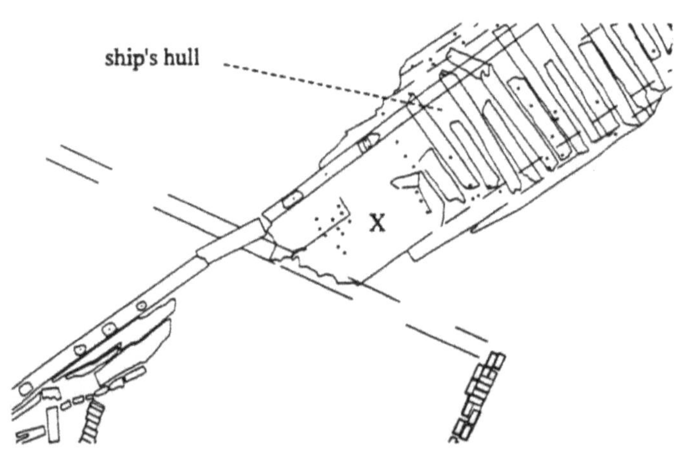


**Figure 29** Position of the square weight (X) within the ship.

**Weights from Building Five and Surrounding Area**

From Building Five and its surrounding area, thirteen weights were recovered during the 1989 and 1990 field seasons. Most of these weights were found in Room 4 of the building. Building Five consists of Room 1, Room 2, Room 3 (preparation area), Room 4 (cook room) and an adjacent yard area with cistern known as Yard 5. Room 1 had very little in it except a plaster floor and a large beam, which most likely was a main ceiling structure. In what has been designated as Room 2, a stairwell was the predominant feature. In the space underneath the stairs, was a storage area in which a minimum of 22 pewter plates were found. The remnants of four brass candlesticks, two large ceramic vessels, one filled with pitch, an ink sander and one weight were recovered in the room. The largest concentrations of weights were found in Room 4 which was a kitchen area with a large hearth. Dome weight PR89 859-12 (A), which weighs 30 pounds minus its staple and iron ring handle, was found along with PR89 857-9 (B) and PR89 859-1 (C) in close association with the hearth in the South-West corner of the room. PR89 857-9 and PR89 859-11 were 25 pound and 55 pound weight respectively, with their staples and iron ring handles. The mass of these weights and their association suggests that they were used in conjunction with fire place machinery. A four ounce lead bun weight, PR89 867-3 (E), was recovered in the North-East corner of Room 4. This weight is a common weight type use on scale pans. Three bronze nested weights, (PR89 964 7-1, 064 7.2 and 964 7.3 (F) were recovered from Yard 7, located adjacent to the cistern. Their close proximity to Room 4, as well as the hearth of Yard 7 may indicate their use in food preparation. Distribution of Building Five and surrounding area weights is represented in Figure 34. A single bronze nested weight (PR90-941)(G), was found in Yard 5. This weight cup did not resemble the three nested weights found near the cistern and this cup measures a mass of 4 ounces Avoirdupois. A single lead bun weight, PR89 867-3 (H), was recovered in Room 2 of Building 5. Circular in shape, this weight has a deeply inscribed "S" on its top and bottom surfaces, which may be owners marks. This weight measures 4 ounces Avoirdupois in mass and would have been used on a small set of scales (Figure 30).

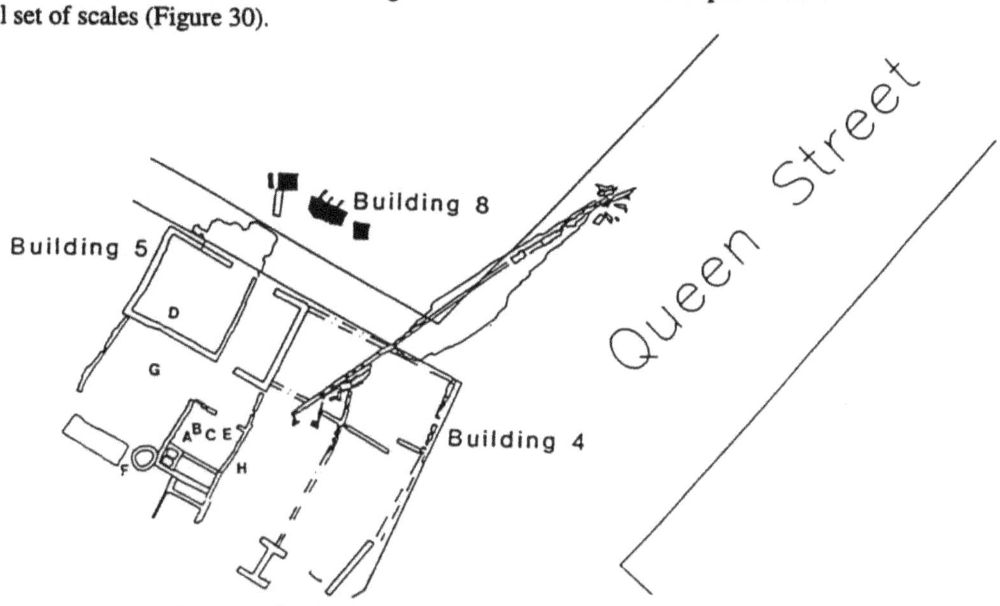


**Figure 30** Site plan for Building Five and surrounding area.

Bronze nested weights account for the remaining 40% of the weight assemblage in this area. Nested weights PR90 964-7.1, 7.2 and 7.3 were recovered and by their conserved weights, it is possible to establish that the complete set of weights undoubtedly weighed two pounds Avoirdupois. On their top rims, these cups are inscribed with the Roman numerals XVI, VIII and IIII, according to their decreasing size. These numerals represent ounce markings of sixteen ounces for the largest cup, eight ounces for the next smaller cup and four ounces for the smallest of the three cups. Each cup has two Troy ciphers and two lions passant markings inscribed on its bottom edge. Generally, the exterior cup of a nested weight set, is equal to the weight of the remaining interior cups. Since the largest of the interior cups weighed .996 pounds (probably a 1 pound cup) after conservation, the remaining smaller interior cups totaled one pound in mass. If, as was the custom, the weight of the exterior cup was to equal the combined weights of the interior cups, the exterior cup would have measured 2 pounds. The nested weight set would have been a 4 pound Avoirdupois nested weight set in total mass. Knowing that most interior weight cups were sized at 50% of the cup that they fit into, projecting the weight of missing weight cups is possible. In the case of PR 90 964-7.1, 964 7.2 and 967 7.3, we can determine that the nested weight set was composed of five cups, ranging in increments of 16oz., 8oz., 4oz., 2oz and 1oz.cups. An additional one ounce disc weight was customarily (but not always) added to complete the set to an exact two pounds. This set of weights is very similar to a set in the Streeter Collection which Ellen Danforth included in her book on nested weights. While Nuremberg was the main center for the manufacture of all nested weights, the lions passant on this particular weight set designates that it was made in accordance with the standards set by the Goldsmiths' Hall in London (Danforth 1954: 97). All three weight cups have recessed bottoms which were created during the lathing and weight adjustment process most commonly used by English craftsmen. A fourth weight cup, PR90 941, was found approximately thirty-five feet from this set. Neither its weight or markings suggest that it was part of the larger nested weight set. Based on mass alone, it is impossible to say what type of nested weight set this cup belongs to, since stylistic interpretation is difficult due to the fact that muck of the outer surfaces details have exfoliated. Consequently, it is hard to tell if this cup weight belonged to either a Troy or Avoirdupois standard set. The bottom surface of this cup is flat in a style similar to weights from the Nuremberg weight makers. Weight cup number three from the Monte Christi nested weight set is similar in size and mass to PR90 941. While it is possible that these weight cups are from similar sets, Henry Skinner, a London manufacturer of scales and weights in 1869 reports that the largest proportion of older weights brought to him for adjustment were substantially lighter than the standard (Crellin & Scott 1969: 59).

In Room Four of Building Five, a large brick weight (PR89 857-9) with some unusual features was found. The weight had the Roman numerals "LV" inscribed on its top surface and minus its iron ring handle and staple fastener, weighed 55.2 pounds after conservation. There are three nail holes in the top left surface of the weight and traces of iron oxide found in each suggest that the weight had been adjusted by hammering iron I and clinching their ends into the lead. If the "LV" mark was added to the weight to indicate 55 pounds as an intended weight, the addition of an iron ring handle and nails made this a heavy weight. On the side of this weight, three "SP" initials have been stamped into the lead. During the 1989 field season, a small lead weight (PR89 867-3) with a deeply incised "S" mark on each flat surface was recovered. A pewter plate from the same area also has the "SP" inscription.

Two square weights of a similar design were recovered in Room 4 of Building Five during the 1989 field season. PR89 859-11 (Figure 31) and 859-13 (Figure 32) are two excellent examples of upright standing brick (trapezoidal shaped) weights. In both cases, well preserved markings have been recorded from these weights. PR89 859-11 has a "VXX" deeply inscribed weight denomination in its side.

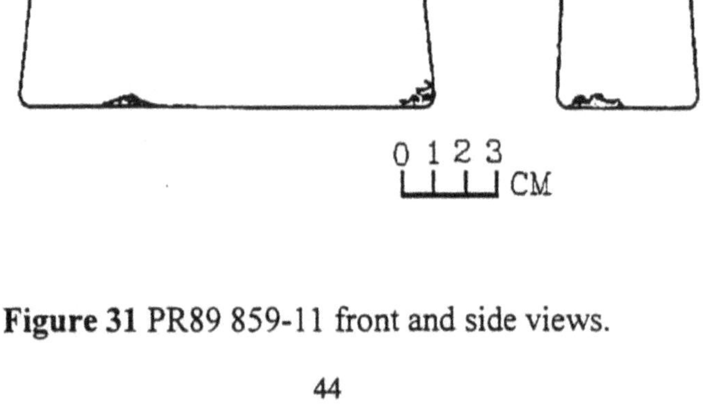


**Figure 31** PR89 859-11 front and side views.

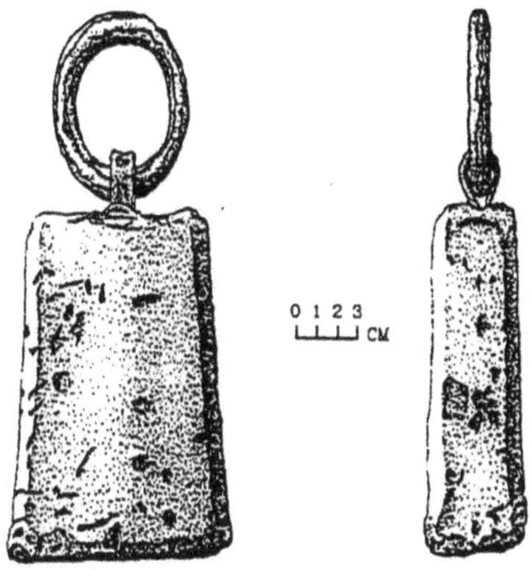


**Figure 32** PR89 859-13 front and side views with iron ring handle and staple.

The presence of staple holes on the top surface of this weight suggest that the weight originally had a small iron ring handle. On each side of the staple hole is a well preserved ship insignia complete with rigging lines. Located above this ship figure in each insignia is a large Fleur-de-Lys mark. The earliest use of this mark is on an English coin weight, dated circa 1650 (Houben 1978: 8)(Figure 33). Figure 34 is a stamp taken from a rubber peel made of the mark on PR89 859-11.

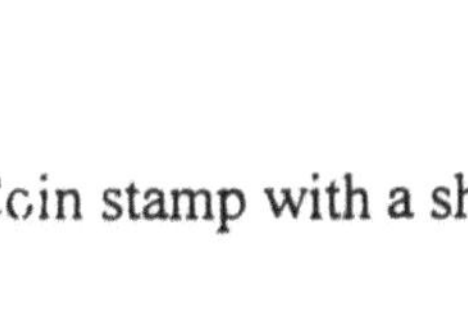

**Figure 33** Coin stamp with a ship and rigging.

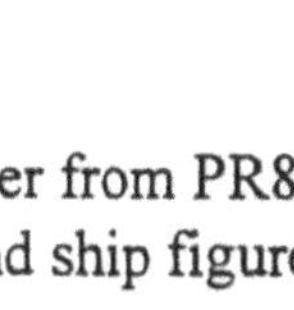

**Figure 34** Cipher from PR89 859-11 with Fleur-de-Lys and ship figure.

After conservation, this artifact weighed 24.75 pounds Avoirdupois. The size of the staple marks on the top surface of this weight suggest that the weight may have had a very small staple. If this was a twenty-five pound weight, as its inscription suggests, then the ring handle must have been very small. Based on the projected weight of the ring handle that was reconstructed for PR89 859-13, the weight may have been slightly heavier than twenty-five pounds. The iron handle of PR89 859-13 had completely disappeared due to the corrosion process of iron in salt water, but a large concretion formed during the process enabled reconstruction of the handle and staple using epoxy casts. This weight has no weight inscriptions on its sides and weighs 13 pounds and 8 ounces, minus the iron handle and fastener. With these iron fittings in place, this was a 14 pound Avoirdupois weight.

In both cases, the large ring handles that would have dominated the top surfaces of these weights served to protect their ciphers from damage due to use-wear and denting. Stamp marks on squared weights tend to survive better than stamps on rounded weights because their flat surfaces allow a deeper striking of the mark. Squared weights tend to show extensive wear on their bottom edges which become rounded over time. The sheer mass of the weight and the softness of the lead used in its construction are prime factors leading to the distortion in shapes of heavier weights.

Composite cone weight PR90 897-9.1 is a very unique weight which appears to have been constructed differently than other similar weights recovered from the site. This weight was recovered from the hearth area of Building Five. Much of the metal surface of this weight was badly deteriorated making the task of removing it from the concretion more difficult. Because of its size, the concretion which encased the weight had been shipped dry to the lab which may have caused the matrix of the concretion to harden more than usual. Tests of the remaining exterior surface of the weight indicate that it was probably copper.

The weight was constructed by forming a cone of copper with a folded seam along one side. Unlike PR 305 and PR83 423-40, however, the bottom of this weight was formed by crimping a bottom plate to the side edges of the open end of the cone of the weight. This plate, possibly made of wood or metal, acted as a cap which gripped the straight bottom edge of the cone base. Radiograms of the weight indicate that a handle had extended from the point of the cone and had been fashioned by passing a knotted wire through the hole in the cone before sand was added to form the body of the weight. PR90 897-9.1 is a unique weight in that it broke convention with the standards set by the London guilds which used lead or bronze to manufacture weights. Construction techniques indicate that this was a finely crafted weight. The seam line along the length of the weight and the precise fitting cap bottom plate suggest that much care was taken in its construction. No marks were retrieved during the conservation process. Because this weight was recovered in the hearth area of Building 5 along with numerous pieces of broken tools, large bolts and a broken candle snuffer, it is possible that the weight and other materials were simply stockpiled as scrap metal by a local smith. Analysis of the contents of the large floor concretion is not complete but this weight could have also been used with as part of a turning spit assembly. Because its weight can only be approximated at 1.22342 pounds, it is more probable that the weight was used in conjunction with a small balance beam arm which was recovered from the same concretion.

An iron scale arm (PR90 897-9.11) measuring 18.6 inches was found, adjacent to the weight in the same concretion. While badly deteriorated, markings in the concretion indicate that both arms of the device were of equal length with a small pointer finger extending above the scale. The ends of the beam are badly fragmented but what remains of the broadened ends suggests that the scale had been equipped to handle scale pans as well as hooks for holding weights and loads. PR90 897-9.1 was probably used in conjunction with this scale.

**Patterns In Weight Type Distribution**

Patterns of weight type distribution from archaeological work conducted at Port Royal are very illuminating. Thirty-one of thirty-eight weights associated with commercial weighing activities from the TAMU/INA excavations adhere to the Avoirdupois standard. Only the nested weights adhere to the Troy standard.

Crellin and Scott have suggested that the use of lead weights was in rapid decline in favor of bronze weights and the emergence of the Founders Company as the most powerful of guilds after 1700 (Crellin & Scott 1969: 53). The beginnings of this trend are apparent in the recovered assemblage of weights from Port Royal. From all excavations at Port Royal, lead weights, including lead bun, lead disc, dome and brick weights account for 56% of the total weights recovered. The large number of bronze weights (35% of the total assemblage) in use as compared to the lead weight assemblage is an indicator of the increasing popularity of bronze weights. Prior to 1650, the percentage of lead weights in use would have been considerably higher and bronze disc weights would have been the predominant second weight form. Figure 35 illustrates the rise in popularity of bronze weights.

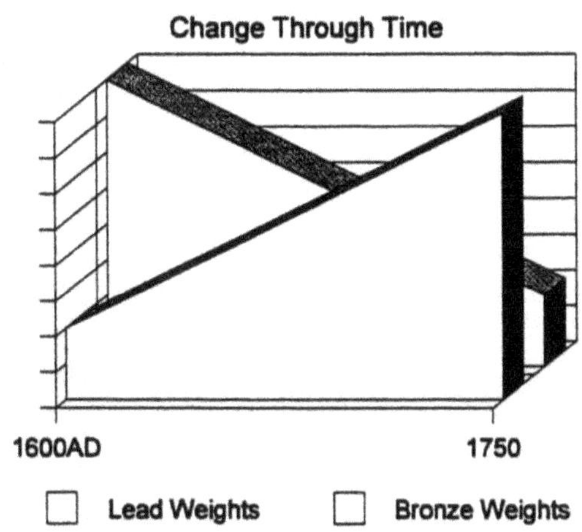


**Figure 35** The rise in popularity of bronze weights through time.

Crellin and Scott devote a great deal of time describing the "laissez-faire" attitude towards enforcement of the laws governing apothecary weights in England (Crellin & Scott 1969: 56). Given that the Westminster Office weighed all types of brass weights including pharmaceutical and disc commercial weights, we can assume that their lax attitude extended to all weight types and not merely the specialized weights used by apothecaries. The trend in overall amount of deviation from the standards indicates that lead weights were closer to the accepted Avoirdupois standard than were bronze weights. This might be explained by several factors. If bronze weights were loosely governed by the Westminster Office, then there was a strong possibility that a large portion of these weight in active use throughout London and district were sub-standard weights. Unless a merchant was to seek out the specialized services of an expert repairman-adjuster such as John Warner of London, bronze weights were seldom, if ever corrected (Crellin & Scott 1969: 54). During the seventeenth century, the Founders Company was the smallest of the guilds with jurisdiction over weights. Their ability to control weight standards was limited until after 1700 when lead weights fell into disfavor because they were softer and prone to heavy use-wear. Because of this, as Crellin and Scott observed with pharmaceutical weights, we can expect bronze weights to deviate from the Exchequer standards more than lead weights. This is true in the case of the Port Royal collection. The majority of weights deviate from the Avoirdupois standard by less than 2%. Only PR85 1034-8 (-2.6%), PR 86 144-6 (+10.5%), PR85 1034-6 (16.2%) and PR85 1034-2 (+10.4%) appear to deviate greatly from the Avoirdupois standard. Including these weights, the average deviation from the Avoirdupois standard for lead weights was 2.5605%.

Bronze disc and nested weights deviated from the Avoirdupois standard by a greater margin of error than did lead weights. Twenty-eight weights or just over ninety percent of the total for bronze weights deviated from the standard by an average of 11.1428%. The remaining ten percent of the bronze weight assemblage was errant from the standard in excess of 11.1428% meaning that bronze weights deviated from the Exchequer weight standard at a rate four times that of lead weights.

With an increase in the production of bronze weights after 1700, the Founders Company of London, which had the authority to stamp Troy and Avoirdupois weights, became increasingly more powerful. Investigations of the weight assemblage from a historic site dated after 1700, therefore, should demonstrate a greater number of bronze disc weights as well as a wider variety of shapes and sizes.

A number of factors indicate that the Founders Company may not have a tight control over the production and standardization of weights before 1700. Several of the bronze disc weights from the Port Royal collection have ewer ciphers with handles facing to the right. This is a departure from the left-facing ewer used before the reign of Charles I. The right-facing ewer insignia is a different style than the insignia used after 1625. PR85 1026-9 and D95-2 are excellent examples of the official ewer stamp of the Founders Company. Weights PR 355 and PR 156 include both have right-facing ewer ciphers. Before 1700, the Founders Company was still a small, fragmented artisans guild with representative members throughout England. Given that the King's control over weight systems through the Exchequer Office was in a state of chaos, it is very probable that guild houses suffered from the same lack of leadership and continuity. Since the reversed ewer was used as a mark for an extended period of time between 1625 and 1685, we might conclude that the change in ewer design was done to mark the reigns of Charles I and II. The insignia may have been reversed in observance of the reign of Charles I or as a means of identifying the office, city of origin or foundry which had issued the weight.

Apart from ciphers and weight types, which one would expect to find in abundance in seventeenth century English holdings, one distinctively different bronze disc weight, weighing one and three quarters ounces, was recovered at Port Royal. Artifact SP.1.S51 was recovered from excavations conducted by Philip Mayes. At first glance, this weight looks like a coin and both surfaces have an "M" inscription on them which has no parallels in the English metrological system.

**The Assize of Wheat and Weight Standards**

From 1685, the standard for wheat had fluctuated between 55 and 59 pounds per bushel, depending on climactic conditions during the growing season (Connor 1995: personal communication). After 1800, this range in weight for a bushel of wheat was still considered to be the accepted standard and a 55 pound bushel of wheat could be baked into 433 pounds Avoirdupois (A.) of wheaten bread as well as 25 pounds A. of household or brown bread, in the Winchester standard (Connor 1987: 204). The assize of 1684 then, accounts for the need for 25 and 55 pound weights for the measurement and baking of bread. The association of PR89 857-9, with its incised mass of 55 pounds and weight PR859-11 (25 pounds) with the hearth area of Room 4 in Building 5 would suggest that these weights were to measure quantities of wheat, possibly for local sale and for on-cite baking.

It is the Assize of Bread of 1710, commonly known as the Act of 8 Anne, that explains the need for 9, 7 and 2 pounds 4 ounce weights which are part of the Port Royal assemblage. Concerned with the standardization of breads and with the just payment that bakers were to receive for their labor, the Act of 8 Anne established weights and prices for all types of bread. In this list, the weight of a sixpence large bread was listed as 2-4-4, meaning 2 pounds, 4 ounces and 4 drams. An eighteen pence loaf of large bread was to weigh 7 pounds, 1 ounce and 12 drams and a twelve pence loaf of household bread was to weigh 9 pounds, 1 ounce and 1 dram (Assize of Bread 1710). Wheat sold for the making of household bread was a more common form of coarsely ground grain used to bake brown bread which sold for 11 shillings and six pence. Grain sold by the seven pound standard was generally more processed and used for the making of a better quality white bread.

Distribution of 7 and 9 pound weights, then, plays an important factor in determining probable commercial activities within Buildings 3 and 5. One seven pound and two nine pound weights were recovered amongst the assemblage of weights in Room 1 of Building 3. In this same area, numerous unsmoked clay pipes and whole onion bottles were also recovered, suggesting that Room 1 may have been used as a vintner's shop or as a storage area. As suggested by the presence of 9 pound A. weights, poorer quality wheat for making household bread, was dispensed from this area. In contrast, as the Acts of 8 Anne of 1710, the Assize of Bread of 1685 and the actual mass of the weights recovered might suggest, finer quality wheat was being processed in Room 4 of Building 5. In this area, 55,

25 and 14 pound weights were recovered. While the 55 and 25 pound weight would have been used to measure quantities of wheat for baking, the 14 pound weight (a multiple of the 7 pound standard) would have been an acceptable weight for measuring eighteen pence loaves, according to the Act of 8 Anne. If Building 5 was in part a tavern, or if the hearth and oven in Room 4 were being used as a bakery, larger incremental standardized weight would have been logical considering larger volumes of baking.

Enormous quantities of wheat were imported into Jamaica from Europe and from the American colonies during the seventeenth century. Bakeries and citizens alike, would have relied on continual shipments of wheat, which would have been off loaded and processed in the community, for local and area re-distribution.

# CHAPTER VI

# THE ARCHANGEL MICHAEL AND THE SIGNIFICANCE OF EARLY CHRISTIAN ICONOGRAPHY IN METROLOGY

For over a thousand years in Western society, the iconography of the cardinal virtues, including all of the elements in the Archangel Michael cipher mark of the Plumbers Guild, have served as markers in written and pictographic records, for social change and variance in interpretation of value systems between and across ethnic and cultural boundaries. From the reign of King John in 1199 AD, attempts at standardizing the English metrological system had been partially stalemated by two opposing forces. Realizing great profits from the use of private weight and measure standards that had been granted them by the Crown, rich land barons and shire towns were reluctant to relinquish their use in favor of centrally controlled standards which the monarchy desperately needed as a means of increasing international and interregional trade. Foreign tradesmen emigrating to England, including artisans making lead and bronze weights, brought new technologies and mental templates to the country. Many of these weightmakers used the stamp of the Archangel Michael as a mark or symbol of quality assurance. Because of its long association with the cardinal virtues, the Archangel figure was also associated with the attributes of trust and egalitarian treatment, which the monarchy held as a central tenant for the development of a monarchically controlled metrological standard. In contrast, German and Italian artisans appear to have interpreted the Michael figure with his sword and scales from a more somber perspective, portraying him as the purveyor of swift justice, who stood in judgement over man and his earthly sins on the final day of reckoning.

### Artisans and Foreign Influences as Mechanisms for Change

At the beginning of the fourteenth century, weightmakers living in the city of London England started to include a variety of ciphers on weight and measures. Apart from the need for marks representing the mass designations of weights, manufacturer's marks, owners' marks and guild ciphers started to appear on weights. By the first half of the seventeenth century, competition for the rapidly expanding merchant markets in London and the influx of skilled foreign craftsmen into the city precipitated an environment in which guilds tightly monitored their royally appointed territories. Monopolistic control over the production and maintenance of weights and scales was the main source of income for the Worshipful Company of Plumbers and the Worshipful Company of Founders of the City of London, which produced weights in small and tightly controlled guild shops. The expanding system of artisans included an ever-increasing proportion of foreigners, and unlike the "freemen of the city," who were either English citizens or foreigners with special dispensation to conduct business within the realm of England, these weightmakers were disadvantaged by royal restrictions which reserved the markets of the city for the freemen. By restricting their ability to conduct business in the city, laws protecting the freemen of London forced foreign artisans to either establish trade partnerships with English artisans or establish interregional and international markets. Historical evidence indicates that foreign artisans within the City of London tapped into both strategies, which ultimately gave them certain trade advantages over English craftsmen. While they were able to gain profits from their partnerships with freemen, filling a need for skilled craftsmen who could make a variety of weights, they also kept a hand on long-distance markets in their homelands. Both English and alien craftsmen enjoyed mutual benefits from their joining forces in the production of weights and measures. While freemen and foreigners managed to coexist and benefit from their shared markets, it appears that social distinctions were rigorously maintained. Most foreigners lived in an isolated part of London named St. Martin's le-Grande. Initially, they did not adhere to the standards set by the guild systems and made weights, measures and scales according to their traditional methods. Established as a parish, St. Martins became a center for numerous trades including the Goldsmiths, Plumbers, Pewterers and Silversmiths to name a few, and these artisans formed their own brotherhoods which kept their traditional craft methods alive (Royal Historical Society 1961: 59).

In competition with the well-established manufacturing and trade network of the Hanseatic League, craftsmen from Flanders, France, Scotland, Scandinavia and Italy emigrated to England, which was fast becoming a formidable center for cultural interaction and international trade activities. Apart from their skills in working with brass and lead, these artisans also brought a wide variety of mental templates concerning the nature of weight and measurement systems to England. These templates included unique manufacturing techniques for weights and scales as well as differing standards for mass and capacity measurement. Although England, Normandy, Flanders, Scandinavia and the Rhineland shared the common weight standard of the mark, which was 216 grams in mass, reliable documentation exists to indicate that there were at least four different weight standards for the English monetary pound between the time of Æthelred II (AD 1018) and the Norman conquest up to AD 1066 (Nightingale 1985: 195). While the Troy standard was reserved for the measurement of jewels, precious metals, liquor, bread and pharmaceuticals, the Avoirdupois standard and the Tower pound systems were used for the measurement of base metals, wool, tallow, hemp and general food items. The Hanseatic system of weights and measures was rapidly adopted by many Englishmen who were forced to compete with the highly competitive, nautically-based trade company. Because of her geographic location, her reliance on the seas for long distance trade and importation of materials and because she was actively seeking new currencies to bolster the British economy, England had become the cultural and commercial crossroads between Roman, Scandinavian, and Moorish influences. Rather than simply compete with other currency and metrological standards, the monarchy reasoned correctly that the establishment of standardized weights, measures and currency in conjunction with her growing appeal as a commercial center would draw other nations to use the English standard for all commercial transactions.

Borrowing Simon Bronner's idea that "material culture describes more than objects," and that artifacts have meanings that "strike deep into the relation of human existence and expression," one would expect that the rich iconography of the Plumbers Guild Archangel figure is steeped in religious meanings which the monarchy utilized as a means of popularizing its new standard (Bronner 1992: 18). Several symbolic images have been combined in this single cipher. When analyzed as a whole or as component parts, the images included in the Archangel Michael cipher of the Plumbers Company are pervasive. The Archangel symbol has been adopted and used in a variety of medieval English contexts and has been adopted by the Church of England, the Tower of London, the Plumbers Company of London and generations of artisans who, over time, have utilized the mark to draw an association with English freemen and the high quality standard associated with English craft guilds.

It should be expected that with her rise as an economic and commercial center, England became a crossroads for complementary and divergent technologies in manufacturing and cultural influences. The image of the Archangel Michael, the sword, scales and the range of clothing associated with the figure, have long histories of symbolic meaning. So too does the walled parish of St. Martin's le-Grande which physically and socially isolated the alien artisan population of London. The cipher of the Worshipful Company of Plumbers of the City of London represents a unique amalgamation of symbolically rich images representing the cultural milieu from which the British metrological standard was derived. Because the Archangel stamp was placed on lead weights, which are prone to use-wear and physical disfigurement due to their great physical mass, many of these ciphers are distorted or partially obliterated. The Plumber's Company stamp seldom survives in perfect condition. Because of this, and because the images suffer physical distortion because of the soft lead matrix which they are stamped into, no single descriptive process can adequately describe the physical properties as well as the artistry involved in the Michael figure. This elaborate cipher has been recovered on two pail weights from excavations at Port Royal, Jamaica. Representing just over 1.44 percent of the 138 weights recovered to date at the site, these weights, which have an absolute mass of 14.0 pounds each, represent a royal mandate for the standardization of English weights and measures. Their presence in England's New World colony is significant, marking the fact that wide sweeping reforms in England's metrological system were directly affecting trans-Atlantic commerce. The inclusion of the Archangel cipher and the elaborate construction techniques used to manufacture them suggest that they were special weights within the English metrological system.

**Review of the Literature and Other Sources of Text**

In order to study the Archangel cipher and understand the significance of the imagery included in its design, it was necessary to utilize a great variety of literary sources as well as maps of the old core area of the city of London. Platts and land records were not readily available for this study, but I was fortunate to locate old tourist maps dated from the mid-nineteenth century. Maps such as these proved to be very helpful because unlike official land platts which, by necessity, needed to be updated regularly, these maps marked points of interest including commons areas and alleys and lanes that had been associated with crafts and old industries. Because these maps were for the purpose of locating areas of interest as opposed to delineating the absolute locations of current roads and districts, they proved to be most useful.

Dictionaries such as *The Oxford Dictionary of the Saints*, *The Dictionary of Medieval Civilization, The Dictionary of the History of Ideas; Studies of Selected Pivotal Ideas* and *The Facts On File Encyclopedia of World Mythology and Legend* were useful for their abundant references. These sources were invaluable because they provided some basic background information about the angel figure. While the compilation of data was helpful in constructing a generalized overview of the Michael figure, the information was lacking in culturally specific details. Liturgical sources including *The New Catholic Encyclopedia, Butler's Lives Of The Saints* and *The International Work Of Reference On The Teachings, History, Organization, And Activities Of The Catholic Church And All Institutions, Religions, Philosophies, And Scientific And Cultural Developments Affecting The Catholic Church From Its Beginnings To The Present* were useful texts. As there titles suggest however, the information taken from these sources was directed and formulated towards an understanding of the saints from the Catholic Church's perspective. While the king's movement towards a new centrally controlled standard would seemingly have little to do with the doctrine of the Catholic Church, it must be remembered that interaction and controversies between the Holy Roman Catholic Church and the Church of England involved much more than wars over fundamental tenets. Derived from largely common beginnings, the popular symbolism of both churches has remained strikingly similar. Additionally, the tenants of the Catholic church were central in the minds of the foreign artisans who formed the technological and artistic base for weight making in London and for the movement towards standardization of the metrological system.

In many respects, I found that the most useful sources of information were the numerous and seemingly obscure references to the Saint Martins's le-Grande that appeared as footnotes in several books which focused on the formation of English guild systems. Considered as discrete pieces of information, these references seemed almost insignificant and they appeared to contribute little in the formation of a concrete picture of either the social organization or guild structure that was centered in the area. Considered cumulatively however, data taken from *Travelling Brothers*, *Wardens' Accounts Of The Founders' Company, 1497 - 1681* and *English Guilds* provided many details about English parishes and their activities in forming guilds and attracting foreign artisans that were necessary in order to better understand the social structure surrounding the lives of non-freemen in a highly stratified society. Small footnotes mentioning St. Martin's le-Grand as a walled precinct most noted for its large population of foreigners (aliens) and disreputable work force, speak volumes about social, economic and artistic interaction.

In many respects, the most rewarding part of the exercise was in attempting to delineate the demographic distribution of various guilds in relationship to the prestigious economic center of London. I found that tourist maps dated at 1866 were very useful in helping me tentatively determine the locations of guild offices based on street and alley names. Knowing that street names often reflected the activities of parish and neighborhood guilds, it was a simple task to locate the likely areas of guild activity, based on street names and general proximity to the known old core of the city. Deeper investigation of historic documents has, in many cases, confirmed these locations to be correct. While the location of all guild centers is by no means complete, plotting the locations of the guild houses in the original core of London has sufficiently demonstrated that St. Martin's le-Grande was a segregated community, in terms of its walled segregation and great distance from the actual core of the commercial community. Information from other text sources verifies that the movement towards standardization of England's metrological systems was in part due to the artistic and technological abilities (the mental templates) of foreigners.

The last, and generally most informative sources of information, were poems and illustrations taken from German, Italian and English murals and paintings. The Archangel Michael figure, which is commonly found in a variety of forms throughout Europe, dates to the earliest days of Western Christianity. For over one thousand years, elements of the Archangel Michael cipher have appeared as representative icons for the cardinal virtues which have been generally accepted as representing the fundamentals of the value system of Western man (Wiener 1973: 378). Pervasive in religious, social, political and personal interactions, these icons hold meanings in Western society that have continued into the twentieth century. Standing victorious over Satan with his sword raised in hand, the Archangel is powerful iconography for the virtues of honesty and equality of justice that we take for granted as underlying the principles of weight and measurement systems. As Dell Upton has observed, the study of the symbolism in works of art constitutes a "primary rather than supportive" type of inquiry (Upton 1985: 71). The paintings of Albrecht Dürer and the sculptures of England's Sir Jacob Epstein differ greatly in their intended messages, but the common denominator of a victorious Michael standing over Satan is universal. In his painting entitled *Die Apokalyptischen Reiter,* Dürer shows great disdain for rampant earthly corruption which he depicted throughout society. This is the central theme in many of his works and he depicts the horsemen of the apocalypse as trampling a representative population of sinners, including the clergy, nobles and common folks, while the Archangel looks on. Alternatively, much of the art of Epstein encapsulates worldly corruption in the more universal figure of Satan, lying prostrate on the ground with Michael hovering victoriously over him. Regardless of the medium, style and focus used for artistic expression, countless art works throughout Europe have included the Archangel as a symbol of upholding the common good of mankind and the triumph of good over evil.

Collectively, these sources were useful in developing a fuller picture of the social and economic environment in both England and Europe that accommodated the influx of foreigners and new kinds of thinking. Social changes always evoke social commentary. It was necessary to utilize a wide range of historical and artistic texts to better envision the social environment in which the Archangel Michael figure became a seemingly universally accepted icon.

Although no date has been accurately established, King Charles I (1625-1649) adopted the Archangel figure for use on lead weights. In Figures 36 and 37, the angel figure is visible on the top surface of the pail weight, in association with the crown over "C" insignia of Charles I and II and the dagger insignia, signifying that London was the city of origin for the manufacturing of the weight. The dagger mark, often referred to as a broadsword, became closely associated with mercantile activities amongst the freemen of London and over time, became a symbol in its own right. Apart from the Archangel cipher, the dagger became associated with guild trades in the city and many artisans from distant towns fraudulently applied the mark to their products, because of its close association with the high standard of craftsmanship.

Two versions of the Angel cipher are known. Although it is sparse in engraved details and more abstract in design, the angel figure in Figure 38 has the large wings and scales associated with the Michael cipher and was used by a family firm known as De Grave and Company. Figure 39 is a more detailed Archangel cipher which has been associated with weights made by the Plumbers guild and the shop of Thomas Overing. While Overing, located at Bartholomew Lane (near the Royal Exchange), used the angel figure with sword and scales on his weights until at least 1620, it is the Plumbers Company of London who is most widely associated with using the stamp (Le Cheminant 1979: 290). The shop of Mssrs De Grave, located in St. Martin's le-Grande district of old London, actively used the insignia on all their weights and they stamped each scale that they made with the insignia. The firm was established in the late seventeenth century and remained a viable company using the angel figure until 1920 (Crellin and Scott 1969: 64 and Le Cheminant 1979: 289).

**Figure 36** Top view of a pail weight.

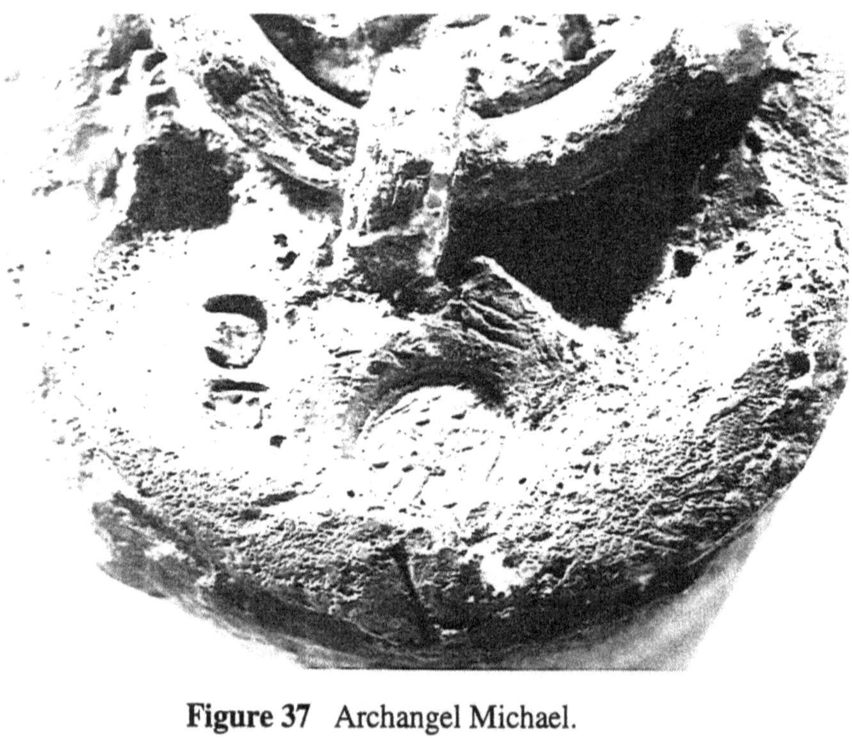

**Figure 37** Archangel Michael.

**Figure 38** The simplest form of the Archangel Michael cipher.

**Figure 39** The Archangel cipher most associated with the Plumbers of London.

**Ste. Martin's le-Grande: Social and Cultural Isolation**

Late Medieval English society revolved around a steadfast affiliation with the church. Parishes depended on the constant support of their entire church family, whose formation was based not only on the sharing of a basic set of church doctrine but other considerations including social and cultural ones. In a similar fashion, guild solidarity was based on selective association rather than geographic proximity (Wright 1988: 35). The church and guilds then, were structured on a strikingly similar basic set of doctrine, but each was different in the nature of their social grouping. While a church congregation could encompass a wide economic cross-section of the city's faithful, parishioners paid their tithes and had no say in how their offerings were spent (Wright 1988: 38). The church tended to draw people of cultural and ethnic similarity together while guilds added the dimension of selectivity in adding members. Wright suggests that the "critical element of choice in participation" set guild membership apart from the less discriminating church structure (Wright 1988: 35). Entrance into the fraternity of a guild system depended, in part, on some degree of economic success. Mandatory attendance and the ability to contribute monetarily to the well-being of the brotherhood was a prime factor for admittance into many fraternities. In this respect, the guild mandate acted as a segregating force against many foreign artisans. It was natural then that churches would draw culturally similar peoples together and that guild fraternities, utilizing the same social demography, were formed as a social sub-set of the church. Guilds attracted parishioners of similar economic status.

The formation of St. Martin's le-Grande dates to the end of the twelfth century. Based on early Saxon ideals, guilds during this period formed the function of serving as the forum for bonds within a turbulent and changing feudal society in which noblemen and commoners were settling into a new metropolitan system. After four centuries of transition, however, Unwin suggests that guilds had changed little in their mandates and structures. He stated that guilds were "providing a social basis for a clique of wealthy merchants bent on monopoly, for a body of journeymen plotting to raise their wages, and for a band of peasants who are being encouraged by their parsons to consult Doomsday Book and cast off all servile obligations to their lord" (Unwin 1908: 53).

St. Martin's le-Grande, or Seynt Martyns Grante[1] as it was commonly known, was a small, relatively poor parish outside the heart of old London. Partially walled, as many parishes were, the parish attracted "aliens" or "forrens" who settled in abundance in the area (Parsloe 1964: 23; Leeson 1979: 52). Situated in relative isolation, West of the core of London, the parish was a conglomeration of foreign artisans, as well as minor government officials, living in the region. In a paper he delivered on the 11th of March, 1961 entitled *The London Goldsmiths Circa 1500*, T. F. Redaway related a great deal of information about the hierarchy amongst guilds in the city, geographic location of the alien populations and their social status within English society. As part of his description of the layout of London guilds he describes;

> "Closer to home, and long a source of dislike, another group of aliens lived within the walls of the precinct of St. Martin le Grande. Members of the Company, but protected from the civil authorities by the charter and liberties of the precinct, the suspicion with which the Company regarded them may be seen in the clerk's triumphant note when, in 1491, one of them came over from the suspect side of the boundary. Bartholomew Naylond, in sum it runs, born in Almain, sworn as a stranger in 1472, having paid £4 in 1490, is admitted a freeman of the craft and a brother of St. Dunstan, being recommended by men of influence, made a denizen of letters patent, and having taken a house in Candlewick Street, wishing to escape from the ill repute of the ill workers of St. Martin's le Grand" (Redaway, p.47).

Figure 40 is a map including the old district of the city. Maps such as this are an invaluable source of information about the social structure and juxtaposition of craft guilds in relationship to parishes and the commercial centers of the city. An examination of the cluster of orange dots on the maps suggests that most prestigious and long-established guilds were centered around Cheapside and Cornhill Streets. In this region, there are several churches and cathedrals which probably acted as centers for guild activities, drawing artisans to the region. Notably, the location of the parish St. Martin's le-Grande (A on the map) (Figure 40) is a great distance from other guild locations. The large dot, labeled Thomas Overing in the central core, marks the location of this Plumber's shop circa 1600. Little is known of Overing to date and it is possible that he was an English weightmaker who was free to establish his shop anywhere in the city. The expense and time required in the printing of maps was high in the seventeenth century, and this map, dated 1677, most likely represents an accurate picture of the parish in the late seventeenth and early eighteenth century. When it was made, there were one hundred and twenty-three parishes with eighty-two chartered fraternal companies. Apart from location, the map is useful in delineating the commercial and craft guild infrastructure within the city. Street and alley names indicate where concentrations of similar craft and manufacturing activities were most likely centered. While it would be inaccurate to conclusively state that as in St. Martin's le-Grande, craft areas were also ethnic enclaves, street and alley names strongly suggest commercial activities that were often delineated in a socially structured hierarchy. Bread Street, Butcher Row, Goldsmith Street, Carpenters Yard, Curryers Alley, Brewers Yard, Three Daggers Alley, Doctors Commons, Rope Makers Alley, Mercers Court, Tenters Alley, Founders Court, Drapers Alley and Brick Layers Line to name a few, received their names as a result of being centers for these trades or companies. The twelve major corporations or companies in London were the Mercers, Grocers, Drapers, Fish-

---

1

Various spellings for St. Martin's le-Grande appear throughout the literature. Because most spellings were not fixed, variations were common and as the name St. Martin's Grante suggests, the precinct was formed as a result of land and grants of jurisdiction, granted to the church (Parsloe 1964: 23 and p.41).

Mongers, Goldsmiths, Skinners, Merchant-Tailors, Haberdashers, Salters, Iron-Mongers, Vintoners and Cloth Workers (Ogilby and Morgan, map explanation). All other companies, including the Worshipful Company of Founders of the City of London, had guild halls and regional administration offices. There is a clear-cut hierarchy amongst the twelve major companies with the Mercers and Grocers being the best established and most influential companies at the time. Physicians, or Physics as they were often known, were also amongst the social elite.

**Key To Figure 40**

| Letter | Identification |
|---|---|
| A | St. Martins le Grande |
| B | Goldsmiths Row & Central Office |
| C | Bread Street |
| D | Leather Lane |
| E | Threadneedle Street |
| F | Fish Street |
| G | Wood Street |
| H | Skinners Row |
| I | Mercers (near Threadneedle) |
| J | Lead Alley |
| K | Beer Street (by King Street) |
| L | Iron Mongers (near King Street) |
| M | Golden Street |
| N | Jewel Street |
| O | Cloth Fair - merchant tailors & drapers |

Figure 40 Map of London, England circa 1840 showing guild centers.

Separated by geographic location, the area in which doctors practiced was named Doctors Commons. This Commons was located outside the crafts district, set apart from the trade-oriented companies located in the old core of the city.

While researching the locations of guild offices and centers in Medieval England, it became apparent that in Bruges and Paris, there was a similar distribution pattern for the location of guild offices and centers. As in London, the twelve major corporations were tightly grouped while less important guilds were situated at greater distances from the core area of the town. Figures 41, 42 and 43 are maps of London, Paris, and Bruges circa 1130 AD. In each map, guilds grouped in close proximity to one another are major corporations, as described by Ogilby and Morgan. Similar distribution patterns for guilds were occurring in identical fashions, in different countries. Table 6 is a comparative chart of the guilds represented in the core areas of all three cities. It is evident that the primary guilds or corporations, which are located within the shaded boundaries of the maps, make up the greatest number of trade guilds represented, suggesting that the concept of guild hierarchy applies in all three areas.

Noticeably, the Founders and Plumbers Guilds were smaller and less prestigious organizations. As mentioned earlier, the monarchy stood alone in its quest to standardize weight and measurement systems. It was not in the interest of land barons and successful shipping companies, who had been granted royal permission to use and maintain errant metrological systems, to abandon systems that had given them an additional potential for economic advantage. The Founder and Plumbers then, represented craft industries of secondary importance. The proliferation of weight systems in England meant that English weightmakers, prior to Henry VII's Ordinance of 1488, were secure in their crafts because from region to region, differences in weight and measurement systems ensured that each craftsman could control a virtual monopoly on the production of weights for a region. Instead of centralized craft or guild houses and craft centers, weight making was a regionally oriented craft.

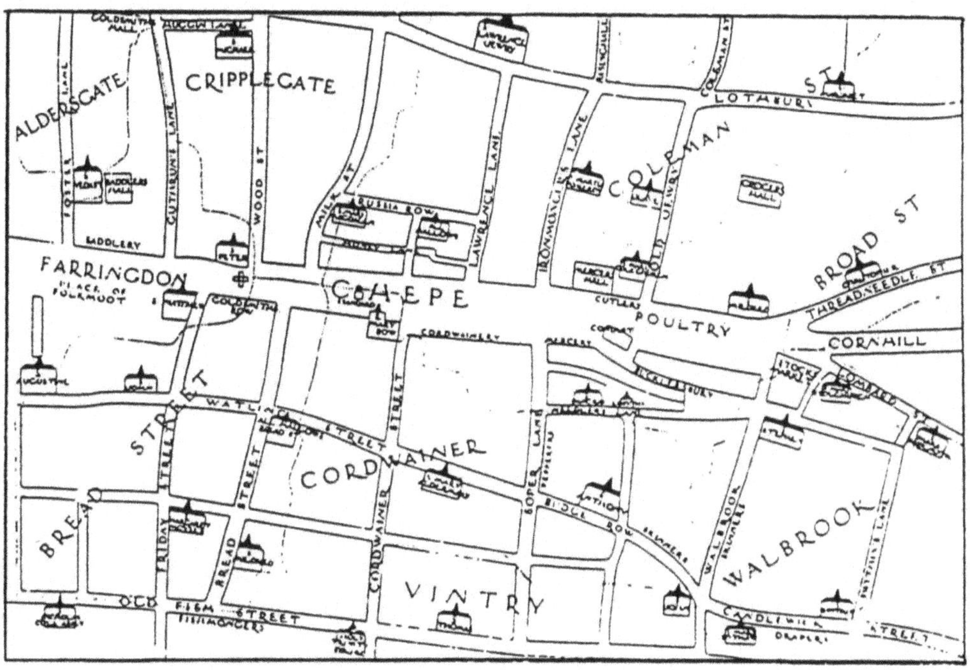


**Figure 41** Map of London, England circa 1130 AD. ( Reprinted by permission of Frank Cass & Co. Ltd, London)

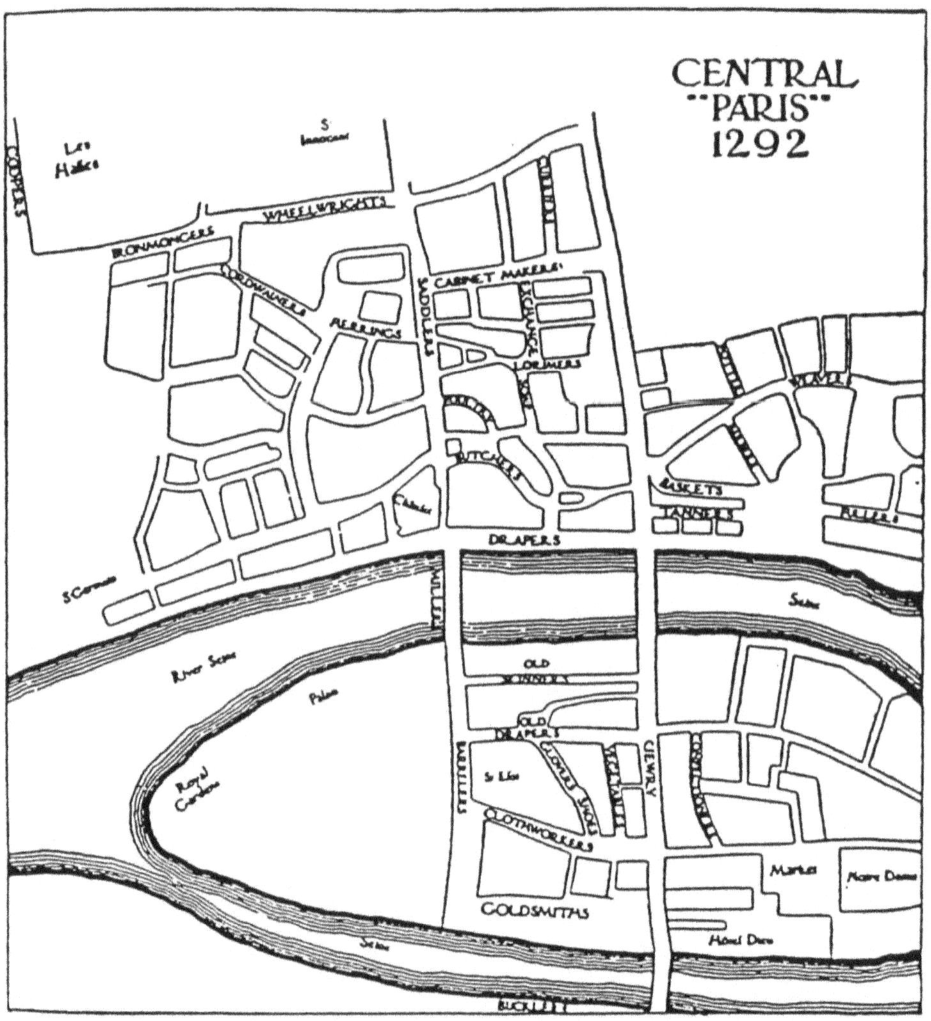


**Figure 42** Map of Paris, France showing street names and guild centers. ( Reprinted by permission of Frank Cass and Co. Ltd, London)

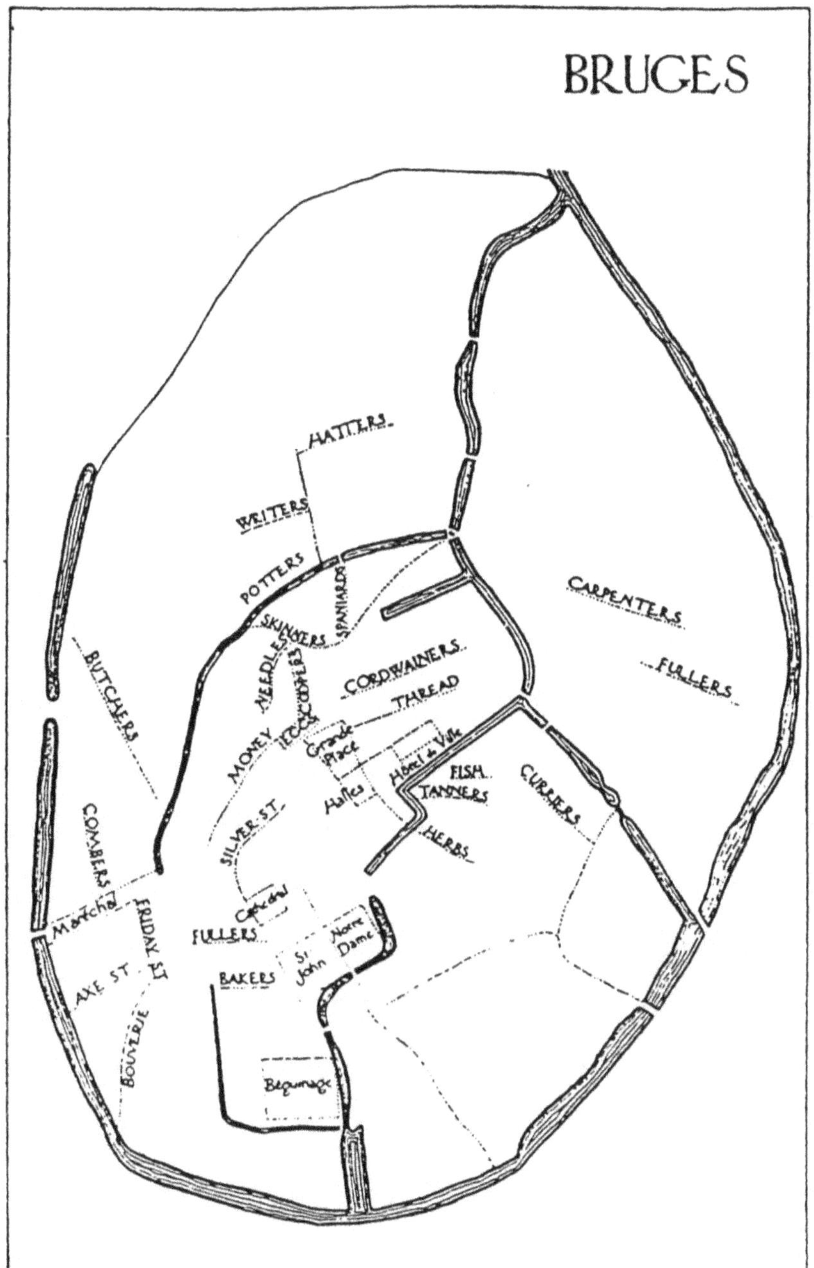


**Figure 43** Map of Bruges, Belgium showing street names and associated guild centers. ( Reprinted by permission of Frank Cass & Co. Ltd, London)

**Table 5 - Key to Figure 43**
**Location Of Main Medieval Guilds In London, Paris and Bruges**

| London | Paris | Bruges |
|---|---|---|
| Goldsmiths | Goldsmiths | - |
| Goldsmiths Hall | | |
| Fishmongers | - | Fish<br>Hotel de Ville |
| Grocers<br>Grocers Hall | Grocers | Grocers |
| Mercers<br>Threadneedle St. | Mercers | Mercers<br>curries, fullers |
| Drapers<br>Candlewich St. | Drapers | Drapers |
| Skinners | Skinners | Skinners |
| Merchant<br>Tailors | - | Merchant<br>Tailors<br>Grande Place |
| Ironmongers | Iron | - |
| Clothworkers | Cloth<br>Weavers | Clothworkers<br>fullers & combers |
| - | - | Haberdashers |

The influx of European and Mediterranean weightmakers into London then, was in response to filling the need for artisans who were not part of the regional systems, to standardize the English system of weights and measures. This had the effect of establishing two opposing motivational forces. In order to capitalize on her status as the foremost commercial and economic center, and create an inward flow of new currencies which she needed, England was forced to rapidly create a standard for weights and measures that would be universally accepted. Pressure to obtain international markets and currencies undermined regional monopolies for the manufacturing of weights that weightmakers had enjoyed throughout Medieval times. It is little wonder then, that foreign artisans were segregated and ultimately forced to seek out trade partnerships and foreign markets as a means of survival.

Alien artisans were viewed with a combination of disdain and admiration. While there was great concern that "the King's naturall subjects would be utterleye decayed" by their presence, alien workers were also viewed as "willing and active artisan members that were easily mobilized through their churches which reinforced their solidarity" (Leeson 1979: 52 and 73). If Englishmen were concerned about the demise of their social structure in the early 1500's, their worst fears had turned to reality by the 1860's. In a series of lectures entitled, "Neighborhood Guilds, An Instrument Of Social Reform," Stanton Coit, a lecturer of the South Place Ethical Society of London, addressed the issue of neighborhood guilds as a means of "lifting the fallen classes of society into independence and prosperity" (Coit 1974: 1).

At this time, the differentiation of aliens from citizens appears to have been insignificant because generally, much of England's populations were economically divided and poverty was pervasive across all class lines. His lectures called for the formation of neighborhood guilds and his basic tenants underscored many of the same social and structural components that had been created in the formation of St. Martens le-Grande and other parishes in the seventeenth century. Notably, Coit called for the fulfillment of the social ideal. This ideal, or mental template, called for the formation of clubs (guilds) which included many of the same mandates that had been sought by the earlier church guilds. Instead of the tight association with the church however, Coit's new mandate called for the formation of guilds that:

> irrespective of religious belief or non-belief, all the people, men, women, and children, in any one street, or any small number of streets, in every working-class district in London, shall be organized into a set of clubs, which are by themselves, or in alliance with those of other neighborhoods, to carry out, or induce others to carry out, all the reforms - domestic, industrial, educational, provident or recreative - which the social ideal demands (Coit 1974: 7).

As Coit's observations suggest, matters of economics were central issues in the mid 1800's and not the old issues of differentiation by class, economics, sex and nationality.

**Liturgical Aspect of the Archangel Figure**

The figure of the Archangel Michael has been a long standing icon in Jewish, Christian and Islamic mythology. Referred to as the one "who is like God" (O'Boyle 1967: 793) and the "chief prince of the heavenly host" (Farmer 1987: 300), the Michael figure is usually portrayed as chief protector of the human race. In the East, Michael evolved as the protector of the sick and infirm (Farmer 1987: 301, O'Boyle 1967: 794, Dahmus 1984: 488). His powers were so strong that he was reputed to be the only figure who could rescue souls in Hell. In England, Michael was adopted as the guardian of armies and the protector of all fighting men. Most notably, the Archangel Michael is the only angel figure that was given liturgical reference before the ninth century. Patrick O'Boyle observes that his cult has grown in steadily in popularity since then (O'Boyle 1967: 794).
The powers attributed to Michael appear to be universal in nature. He is often envisioned in apocalyptic circumstances and the Hebrew faith regards him as the strength and comfort of the people of Israel. Armed with a sword and scales, Michael represents victory over Satan who is usually pictured as a dragon or horned, cherub type figure. Envisioned as the merciful emissary of God, Michael was popularly rendered as presiding over the judgement of the willow twigs on the last day. On this day, each twig, which represents a human soul, is brought back from the dead for awards and judgement (Thurston & Attwater 1962: 678, Farmer 1987: 300).

The Archangel Michael figure was revered early in Western Christianity because he appeared as an apparition on three different occasions and demonstrated compassion for humanity and a desire for peace and religion among all men. In the fifth century, Michael appeared on Mount Galgano in Italy. He ordered the people of the region to build a church around three altars which he had laid down. In the sixth century, he demonstrated his compassion for human suffering by ending a pestilence in Rome. Constantine was moved by Michael's compassion and power and he constructed a church at Sosthenion in his honor (Thurston & Attwater 1962: 679, Farmer 1987: 301). This church was named Michaelion. Michael appeared before Archbishop Aubert of Avranches in Normandy where he instructed him to build a large church in his honor (Mercatante 1988: 452). The cult of the Archangel Michael was further popularized by monks and abbots who came to establish churches in England after the Norman Conquest. This popularity is evidenced by the numerous paintings, sculptures, and frescoes of the angel which appeared prior to 1400 AD.

The most popular images of the angel figure show him either standing with sword in hand, victorious over Satan or as the compassionate figure weighing the goodness of human's souls. Both images represent important concerns of the church and both were popular themes in Medieval art. Michael has been portrayed with and without wings and often, as God's mediator on earth, he is wearing a crown. He is often holding the scales of judgement in one hand and many early images show him holding a sword and chains, standing victorious over a defeated Satan figure. Certainly, the Michael figure represented the triumph of good over evil, and the pursuit of Christian principles.

Historically, weights and measures have always been entrusted into the hands of religious and political leaders who have utilized the Michael figure and theme of the triumph of good over evil. Because of this, weights and measures are often regarded with a sober, pious respect. Arguably, the theme of weighing the good in mens' souls on judgement day is the most commonly depicted theme which includes the Archangel Michael. Whatever his origin, the Archangel Michael came to represent the virtues of compassion and triumph over evil that was the cornerstone of Christianity in Medieval and Renaissance Europe.

The angel has been portrayed as both male and female, and while the figure is most commonly clothed in hose, it is occasionally dressed in a full skirt. It is possible that in the Angel-with-scales cipher, the Michael figure is clothed in body armor which, according to style of the day, was similar in fashion to what is depicted in the cipher.

## The Adoption of Strong Imagery For Sanctioned Weights

Regardless of liturgical perspective, the depiction of the Archangel as either the "protector of the sick and infirm" or the "one who is like God," is the same imagery and power attributed to the king of the realm. It seems totally appropriate that the king of England would adopt these long-held images of divine intervention and godliness and use them for the marking of the official weights of the realm. Because earlier monarchs had granted rich landlords special permission to use weight and measurement systems that differed from the official standard, it became necessary to appeal to a higher power to bring these rich land barons into line, and persuade them to abandon their legal right to use non-conforming weights. The munificent storehouse of mental images entwined in the Michael figure were the best images to employ when trying to placate the fears of an uneasy mercantile society that was in the throes of an awkward three hundred year transition to a more standard system of weights and measures.

## The Making of Pail Weights and Aspects of the Cipher

From the profile of the pail weight, oxides along the vertical sides of the weight suggest a great deal about the weight's construction. To form the body of this weight, molten lead was poured into a cast iron pail which when cooled, formed the sides and bottom of the weight. While the lead is molten, a staple with splayed ends, affixed to an iron ring handle, was positioned on top of the weight and as the lead solidified, the handle and fastener were firmly attached to the weight. The top surface of the weight is flat measuring 8.35 centimeters in diameter. Incised in the top surface are several important cipher marks including a crown over "C" mark of King Charles I, a dagger representing the City of London, and the Archangel Michael cipher of the Plumbers Guild of London. These stamped marks are incised to a depth of three millimeters on the top surface of the lead cap.

Only the top surface of this weight was exposed lead and any adjustments to the mass of the weight would have been made by the addition or subtraction of lead from this surface. A visible slag of lead on the surface of PR85 1026-13 suggests that this weight was adjusted to bring it to standard.

Because the staple and iron ring handle are so large, they act to offer protection for the upper surface of the weight, which explains why the marks stamped into the top surface of pail weights are generally better preserved than marks stamped on weights which do not have the ring and staple assembly. In some incidences however, the weight of a ring handle can act to obliterate underlying marks as the handle continuously compresses the soft lead.

A second reason for the clarity of marks which escape destruction from the handle, can also be attributed to the flat design of the weight. Most other weight forms utilize a rounded shoulder design for the top of the weight. It is not as easy to stamp a flat cipher mark into a rounded surface and because of this, there is a variance in the depth of the stamp mark across the diameter of the mark. The central portion of the cipher is cut deeply into the surface of the lead while the outer edge of the cipher is a shallow depression. Figure 44 illustrates the effects of stamping a flat mark onto a rounded surface. Because of the softness of the lead, the shallow areas of the cipher are more prone to disfigurement and obliteration due to use-wear. The high rate of recovery of Archangel cipher marks from pail weights can be directly attributed to their flat surface design.

The cipher on pail weight PR1026-13 from the Port Royal assemblage is ovoid in shape, due in part to what appears to have been a heavy blow to the side of the weight. In Figure 45, which is a top view of the weight, the edge adjacent to the cipher mark appears to be cracked and crazed, and the lead cap has been crushed slightly. The mark measures 1.5 centimeters at its smallest diameter and 1.7 centimeters along its widest diameter (about the size of a dime) and the perimeter of the cipher is delineated by a single incised line. The angel figure fills the interior of the stamp and although some of the fine details of the mark are not visible, the important diagnostic features have not been marred. The angel is carrying a scale in his right hand and the three strand construction of each suspended balance pan is evident. Although the left arm of the angel is obscured by a wing, the angel figure is holding an upraised sword in his left hand.

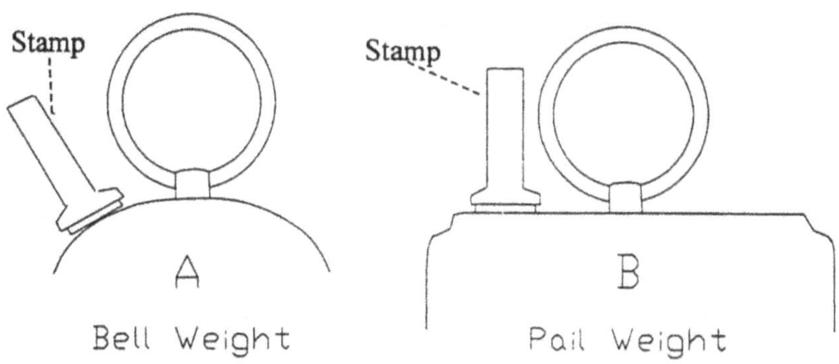


**Figure 44** The nature of stamping on flat and rounded surfaces.

**Figure 45** Top view of the Michael figure.

The addition of a metal jacket around the body of the weight indicates a great deal about the intended uses and importance of pail weights. Apart from making the weights aesthetically pleasing, the metal jacket also set the weight apart as unique from other weights, which is precisely what the monarchy desired. The metal jacket also acted to give the weights a uniformity in appearance that would make it difficult for anyone to either counterfeit or alter, without changing the appearance of the weight. Compound weights, made by combining two or more metals in the manufacturing process, are common throughout Scandinavia prior to 1500, but their presence in England circa 1600 represents a new and a more expensive technology than had previously been used for common weights.

**Turbulent Times and English Economic Expansion**

Apart from the conservation of these weights, extensive background research on the history of social and economic interaction in England were necessary to understand the significance of the use of the Archangel insignia.

Pail weights represent a different kind of manufacturing process than had been used to make all other types of weight recovered from excavations at Port Royal. The use of a cast iron body casing was also utilitarian in nature. The iron casing would have given the weight a uniform appearance and would have acted to protect the body of the weight from nicks, scratches and gouges that would alter the mass of the weight. The elaborate casing of the weight represents a great deal of additional labor and expense in the production of this weight form. Be it by design or by accident, the flat surface of the weights has preserved the Archangel figure, the crown over "C" insignia of Charles I and the dagger representing the City of London very well. Weights of the same mass designation were known to have been used at the central market scales in all of England. Known as the King's Beam, these scales utilized only Exchequer-approved weights and it has been well documented that the Plumbers Company assayed and checked these weights at regular intervals. Examination of the laws and records of Jamaica indicates that, while no guild representatives were sent to the New World to adjust and oversee the use of weights, a market beam, similar to the King's Beam in London, was used in Port Royal. As in London, an official was charged with the duties of regulating and adjusting this scale at all times. Bearing the mark of Charles I &II and weighing a proper designated mass, it is doubtful that pail weights and other weights bearing official stamps were made in the New World. Most weights recovered from the site adhere exactly or extremely closely to the established British system.

During the turbulent time of English economic expansion, Michael was adopted as the guardian of armies and all fighting men. With his sword raised high, Michael was a fitting image for King Charles, as the head of the Church of England, to adopt. The dagger, which is visible along side the Archangel figure on the top of pail weights, became another insignia of the City of Freemen. Most probably, the dagger or broadsword insignia is a synthesis of the association between Michael as the protector of all fighting Englishmen, and the quest for English economic and naval supremacy.

**Arts and the Cardinal Virtues**

In the Middle Ages, issues of temperance and the move to advocating a monastic lifestyle became popular themes in classical doctrine. Works by Martianus Capella, Isadore of Seville and Martianus of Braga's writing entitled *Formula Vitae Honestae*, were a few of the influential literary sources that can be credited for transmitting these classical views. To understand the power of the St. Michael iconography, it is important to trace the evolutionary uses of some of the components found within the cipher. Elements of the icon such as the sword, scales, armored angel figure and crown have separate and independent iconic lives within European art work which can be traced to themes of cardinal virtues (Wiener 1973: 371). Each of the virtues was represented in reliefs and manuscripts by icons associated with attributes of the virtues. At the Palace of Aachen, in Greece, an elaborate plaque showing a tree rooted in a globe includes icons hanging from its branches, representing the cardinal virtues. Prudence or Prudentia, was represented by a book; Fortitudo by either a dagger, shield or helmet;

and Institia by a sword, palm-branch, balance scales or a crown. In numerous manuscripts of the ninth century, Prudentia is commonly portrayed as an open book while Fortidudo is represented by arms and armor. The icon of a set of scales was popular and fixed in its association with the Institia attribute.

By the late eleventh and early twelfth centuries, innovation and elaboration of liturgical iconography was occurring. Virtues were used in a wider variety of devotional and theological manuscripts and new symbolic objects were included in association with them. The Prudence book figure often included a serpent entwined in the pages or a dove hovering over the book and the Justice icon included a figure holding a sword, plumb line or set square. Common to the old Romanesque style of depiction of virtues, in which a standing figure upheld a symbolic representation of a virtue, a relief at the Gothic Cathedral of Chartres demonstrates other interesting symbolic gestures. In this relief, vices are represented as humans in characteristic dress which are being trampled underfoot. Philip Wiener suggests that this representation of uplifting the virtues and the trampling of vices is a common representation style in fourteenth-century Italian virtue cycles (Wiener 1973: 373). Certain icons were universal in their appeal and usage. Twelfth-century French manuscripts include two serpents to represent Prudence, scales for Justice and a dagger representing Fortitudo. A fresco in the Plazzo Pubblico at Siena, Italy, entitled *Good and Bad Government* by Ambrogio Lorenzetti, is a complex amalgamation of icons of virtues and sins. Justice in this relief is represented by an upright sword, similar in style to the Michael figure. This justice figure also includes a crown and a severed head. The Fortitude icon also includes a sword while the Prudence figure points to an inscription rather than the familiar image of him holding an opened book.

By the late fourteenth century, French sculptors adopted the style of including the emblems of the virtues on the tomb of the Duke of Brittany in Nantes and other nobles. Included in these representations of the virtues, Prudence was holding a compass and a mirror, and Fortitude is depicted as holding the neck of a dragon figure which is emerging from a tower which she is holding in her other hand. In these virtue representations, only Justice, represented by a set of scales and a sword, remains unchanged and similar to the Justice icons from Italy, Southern France, Spain and Germany.

The themes of unbending justice expressed in the works of the German artists Dürer and Weigel are foreboding and less optimistic in their egalitarian outlook as compared to the uplifting images of the cardinal virtues which were popular elsewhere in Europe. Dürer's work entitled *Die Apokalyptischen* (Figure 46) clearly demonstrates the symbolic gesture of the trampling of vices that Wiener had observed in the relief at the Cathedral of Chartres. Based on the range of costume styles depicted, Dürer has represented the clergy, nobility and commoners of society in this work. The Archangel Michael is unique in this relief, however, because instead of carrying the scales and sword, he is accompanying the riders of the apocalypse who bear them. Under his supervision they are providing swift and egalitarian justice against the sins of man. Suggesting his mediating influence between heaven and earth as God's emissary, the Angel figure appears to be descending from the clouds which are opening around him. The light of heaven and the grace of God, as suggested by the extended rays behind him, suggest the divine nature of Michael's endeavor. Significantly, he appears to be extending his left hand, possibly touching the blade of the upraised sword which further suggests that this punishment for sins on earth is divinely mediated. The rider figures themselves are interesting because each represents a different virtue. The headdress of the riders with the bow and sword suggest that they are Eastern and possibly oriental in origin, which Dürer has used to represent the virtues of justice and equality. Although the image of the bowman has not been traced in conjunction with the cardinal virtues, the association of "swift as an arrow" is fitting for the scene being depicted in which the riders are quick to trample the injustices of mankind. The bowman is often described as Death and figuratively, the image of the trueness of the flight of an arrow is also fitting in conjunction with the punishing of earthly sin. The rider holding the scales is Justice and he appears to be the strongest of Dürer's characters. The manner in which the scales trail behind him suggest great swiftness in the scene and it seems fitting that the justice figure is the most forceful figure in the scene. Though not pronounced in the relief, the scales represent an advanced form of measuring device in which the pointer or indicator arm passes between two upright supports. Scales such as these were reported to be more accurate and absolute in their measurement since the line of the support also acted to differentiate heavy from light loads. The last of the four figures is an emaciated rider on an equally haggard horse. Traditionally, he has been identified as the Conqueror. His haggard appearance is a symbolic representation of the virtues of steadfastness of conviction required by the virtuous to defeat evil. Instead of carrying one of the icons of virtue, the Conqueror carries a three-pronged pitch fork, which is popularly depicted as the preferred instrument for defeating Satan, the fallen angel. Riding amongst the trampled sinners instead of over them as the other three virtue figures appear to be doing, the Conqueror represents the descending of sinners from grace into eternal damnation and punishment. There is no doubt that the themes of many of Albrecht Dürer's works, as in the relief at the Cathedral of Chartres, cast an ominous prediction on the fate of mankind and the fact that even the Archangel Michael will not intercede on man's behalf.

More philosophical in nature, the painting and poem of T. Weigel expresses similar sentiments (Figure 47). Based around the theme of a weightmaker's shop, the painting includes several types of scales hanging from the rafters. The weightmaker, who is seated at awork bench is looking on as a noble or gentleman figure, as indicated by his cape and hat, manipulates a small hand-held scale. Clearly, one of the pans of the scale is resting higher than the other and the weightmaker's right hand appears to be gesturing about the fact. A translation of a seventeenth-century poem entitled "The Last Day" suggests that possibly, the gentleman figure holding the scale is concerned about the accuracy of the weight and the weightmaker is signifying that only in time will we know where the scales of our lives come to rest. Translated from German, the poem reads:

> How long will you in vanity oh soul
> Here and there totter?
> Weight the scales of this time of lofty thoughts
> And let in all your things (affairs)
> That eternity will tell the final outcome.

A relief of King Henry IV by Tilman Reimenschneider (Figure 48) is an interesting variation on the Michael figure. Differing from the previous Archangel depictions, Michael himself is carrying an upraised sword in his right hand and scales in his left hand. On one pan of his scale is a chalice and a scarab which represents the life of a departed soul. In the other scale pan is a load weight and beneath this pan, the partially obscured figure of demonic figures, which represent the sins and evil of the departed person's life, are acting to pull on the pan and unbalance the soul in judgement. In this relief, Michael has curly hair and large segmented wings similar to the Michael figure depicted in *Die Apokalyptischen*. Carrying scales similar to those held by the apocalyptic rider, Michael is depicted as standing in judgement, possibly of the person who is to his right.

**Figure 46** Die Apokalyptischen (Courtesy of the Fogg Art Museum, Harvard University Art Museums)

**Figure 47** The Last Day, by T. Weigel (After Kisch, *Scales and Weights: a Historical Outline*, Yale Univeristy Press)

**Figure 48** Tilman Reimenschneider's relief of King Henry IV (After Kisch, *Scales and Weights: a Historical Outline*, Yale University Press)

A relief by Adam Kraft from Nuremberg reflects the somber theme of weighing the sins of earthly life and the fate of man. Instead of the Michael figure holding the scales, they are being lifted up by a skeleton featured death figure (Figure 49). In this relief, the scales of justice are not mediated by God's emissary but by a death figure who cannot intercede as Michael is often depicted as doing. This is a tragic scene because the departed soul turns away dejectedly as his life's endeavors are measured. As the pointer arm above the head of the death figure indicates, his sins have proved heavier than his virtues.

Two other examples of art work are interesting in demonstrating aspects of the Archangel iconography. The first is a 13th century Psalter from Southern Germany depicting Michael treading and standing over a defeated Satan figure, depicted as a dragon in Figure 50. Instead of the customary sword and scales, Michael is depicted with an orb and cross-staff, associated him as a princely emissary and holy representative of the church. The halo and large segmented wings are recurring themes in Archangel depictions and were popular because they reinforced the idea of mankind being protected by a heavenly emissary.

Figure 51 is a final illustration of the Michael figure which demonstrates some less common aspects of the traits associated with the angel. As stated earlier, Michael was known as the only angel capable of rescuing souls in Hell and combined in this final relief are several icons that were used in numerous reliefs (Mercatante 1988: 452). As in other depictions, Michael has large segmented wings, a sword in his right hand and a large set of indicator scales in his left hand. Interestingly, the tip of his sword appears to be pointed at, and possibly be touching, a smaller angel figure who is rescuing a human from the scale pan. More visible than in Tilman Reimenschneider's relief, a small winged dragon-like figure is standing on the load pan, ostensibly making the weight of the departed person's sins heavier. Similar figures and demonic beasts with horns, big ears, large noses and tails are depicted as torturing sinners whose souls have not been saved. The expression on the Archangel's face is that of obvious displeasure, possibly because he could not intercede to save some souls, and watched as an agent of Satan tried to weigh the scales in the hopes of claiming a soul in Hell.

**Icons and Their Social Ramifications**

As varied as these illustrations are, all of them are representative icons for the cardinal virtues. Many of the reliefs and paintings associated with German artists revolve around the themes of man's accountability on the final day while others emphasize the compassion of Michael. Underlying either philosophical argument is the common theme of the impartial and swift nature of final judgement. In a time when Englishmen were vying for regional and international trade markets, the image of Michael as the compassionate earthly representative of God was an appealing and universal icon for the De Grave brothers to adopt. The use of the Archangel figure as their stamp dates prior to 1750. The image of Michael as the compassionate emissary appears to have been prevalent in England, as opposed to the foreordained plight of man, that was portrayed in German and Italian works of art. While the Archangel figure used by the De Grave brothers was a simple variation of the complex, sculptured cipher adopted by the Plumbers of London, it contains unmistakable elements to signify that it is the Michael figure, including the large segmented wings and the set of scales in its left hand. This cipher is similar to numerous German representations in that it does not have a sword in its right hand. Unlike these Michael figures however, the English Archangel Michael figure is portrayed as being mighty and assertive as opposed to judgmental. The sword-bearing figure adopted by the Worshipful Company of Plumbers of the City of London has become the form of Archangel figure most often associated with Michael. So strong are the images of the dagger and broadsword that they became symbols for London, commonly known as the City of Freemen.

**Figure 49** Scene over the town gates at Nuremburg, Germany (After Kisch, Scales and Weights: a Historical Outline, Yale University Press.)

**Figure 50** 13th Century Psalter from Southern Germany (Walter Art Gallery, Baltimore).

**Figure 51** Michael the Archangel (After Dürer)

# CHAPTER VII

# PORT ROYAL; A CENTER FOR CHANGE

As a thriving town in a new and expanding agricultural colony, Port Royal was more than a stronghold for pirates and buccaneers. Because she was an economic center and focal point for shipping and receiving for nearly all staple goods coming to the island, the town was a vital link for intra and inter-island exchange. There can be no doubt that the history of Port Royal must, to a large degree, revolve around trading activities and the development of a mercantile economy. In a letter to John Taylor of the Merchant's Royal Company of London, Governor Hender Molesworth advised Taylor that Port Royal was the "most commanding place for profit and advantages on this island" (Pawson & Buisseret 1975: 63). Because of her relative isolation in the Caribbean, early entrepreneurship relied on development of successful merchant activities as well as ownership of plantations, the establishment of long-distance trade associations with London based firms and trade and plunder relationships with Spanish colonies.

With regards to legally sanctioned trade, England supplied nearly all commodities to the original settlement at Port Royal, packaging goods and materials according to the laws and standards of the day. Goods were shipped from England in a wide variety of barrels, butts, casks, hogsheads, puncheons and tierces (Pawson & Buisseret 1975: 69). With a strong mercantile system in place, Port Royal became the most opulent of towns in the New World and yet as Pawson and Buisseret have noted, regardless of her seeming wealth, the town was only a small town by European standards (Pawson and Buisseret 1975: 80).

Day-to-day activities within Port Royal must have appeared to be life in the fast lane, compared to economic growth and development of the rural plantation economy. Port Royal merchants had to know the natures of local and long distance markets and establish business connections beyond the crowded confines of the town itself. To be successful, merchants and businessmen had to rely on instinct, forethought and a sense of good timing. Because of the time lag between the commercial centers of Europe and eager markets in the Caribbean, aggressive pursuit of profit was the only way to succeed. A merchant failing to arrive first at port with his goods ran the risk of trying to unload goods in a satiated market.

In the earliest days of British occupation after 1655, sex ratios in the city were heavily weighted towards a male dominated population consisting of a large military contingent as well as government officials and planters who came in pursuit of riches and resources. From these early days, wholesale adoption of the English system of government and mercantile activities, including weights and measures, was accepted. By 1692, the town had some two thousand dwellings and a population of approximately 6,500 inhabitants. Free men, women and indentured servants accounted for approximately 4,500 of the town's inhabitants with slaves making up the rest of the population. By 1692, Port Royal had a population that demanded finery and commodities that were fashionable in London. Many of the expanding merchant class of the town dressed in Turkish fashion, which was a popularized style of dress patterned after that of King Charles II (Pawson & Buisseret 1975: 109). Indeed, Port Royal followed the trends of Europe very closely and in a letter to his London associates, Thomas Brailsford advised his merchant associates to be on the lookout for "new and pretty things cheap and good for men's and women's apparell, yet newest gymps[10] in fashion and silver footes" (Pawson & Buisseret 1975: 109). Port Royal did enjoy access to all the finery associated with European society. Controlled excavations at Port Royal by the INA/TAMU team under the leadership of Dr. D.L. Hamilton have shed new light on many aspects of life in a British colonial settlement. From extensive archival research, we know that the merchant class of the town was vital to the economy of England as well as the local and regional economy. During this time period, the majority of gold and silver shipments destined for England were shipped through Port Royal. Excavation of Building One has demonstrated the extent of mercantile activity as it was described by Pawson and Buisseret. Combination merchant/shop-dwellings, like Building One and possibly Building Five, were more common than single retail establishments in the community. Dr. Hamilton's research, to date, has also demonstrated that Port

[10] Gymps were silk, worsted or cotton twist fabrics.

Royalists followed traditional English style of construction, essentially recreating a "little England" on the sand spit of Kingston Harbor (Hamilton, personal communication).

The laws of Jamaica and the infrastructure of offices established for day-to-day activities in the town also echoed strong ties to England. The Clerk of the Market assumed the same duties and responsibilities as his English counterpart, the Master of the weigh-house. It can be argued that most probably, the regulation of weights and measures in Jamaica was more stringent than in England. Having to compensate for her colorful and widespread reputation as the "Sodom of the Indies," Jamaican legislation had to enforce strict regulation of weights and beams at all wharfs and markets as a means of attracting honest men and capital investors to the island (Dunn 1972: 149). Because the Clerk of the Market had sole control over the regulation of weights in the town, instead of the convoluted system of guild controls used in England, it is possible that weights in the colony were more uniform in weight than those in England. It is, however, beyond the capability of this current research to determine if this is so.

There is no question that as in England, lead weights appear to have adhered to standards more closely than bronze weights. As has been suggested earlier, this is due to the fact that the Plumbers Company of London was a large and well established guild producing weights that found their way to use in Port Royal. It wasn't until after 1700 that the Founders Company of London gained power with the increasing popularity of bronze disc weights. In the wake of this increase in demand, bronze weight tended to be errant from the standard, more than their lead counterparts. Analysis of the archaeological data suggests that this is true.

From the institutional perspective, the weight assemblage of Port Royal hold no surprises. Evidence taken from the archaeological record points to the fact that as in England, weights were used, re-used and adapted to adhere to changing standards. The founding of Port Royal appears to have coincided with a minor revolution in the regulation of metrological systems and their common use within all British settlements. Inventories from the period of 1680 to 1710 indicate that numerous households had steelyards as well as assorted weights and scales. Before legislation curtailed their use, it is most probable that merchants and citizens alike had used these devices for the dispensing of drugs, cooking and all commercial activities. Sub-section CCXL of the Laws of Jamaica (dated at 1711) brought an end to the sanctioned use of unlicenced weights and beams stating that no "Still-Yards" were to be used in any of the markets, retail stores or hucksters shops (Corporation Of The City Of London, 1698).

In contrast to the institutionalized use and control of weights in Port Royal which appears to closely mirror the British system, the seeming lack of illicit and unconventional weights is more surprising. In reality, the weights recovered to date may only represent a small percentage of the weights present in seventeenth century Port Royal. Excavations by INA/TAMU have been conducted in a commercially developed part of the community and it is possible that future excavations will be conducted in areas representative of different commercial and social activities. Given the closed nature of the city and a generous reward system offered for turning in merchants involved in improper commercial transactions, it seems unlikely that any large quantities of illegal weights should be recovered.
Since Port Royal was a stronghold for buccaneering activities conducted against the entire Spanish mainland and her fleets, it is also surprising that more Spanish weights have not been identified. Weight SP.1.S51, which has been identified as a bronze coin/disc weight, differs radically in weight and style from English weights recovered from the site.

# CHAPTER VIII

## OBSERVATIONS AND CONCLUSIONS

It is obvious that since the extent of the sunken city of Port Royal has not been excavated and analyzed, it is most probable that our collection of weights does not accurately represent all weight types that were in use in 1692. Inaccurate and illegal weights represent less that one percent of the total weights recovered to date. This is an unusually low percentage of illegal weights, considering that most scales and beams were poorly maintained and, therefore, possibly inaccurate beyond acceptable limits set by the Town Clerk for the Market Beam. Excavations have proceeded in an area of the port community which, according to structure types and the wealth of artifact distribution, appears to have been occupied by economically well established shop and tavern merchants. Future excavations in poorer neighborhoods where the artifact assemblages represent non-commercial activities, may well uncover a different assemblage of weights in which the percentage of illegal weights and scales is different. Based on the fact that the majority of the 90 weights recovered to date adhere to the legal standard set by the king and that many of these weights show signs of having been adjusted to bring them to an acceptable standard used at the Common Market Beam, it appears that commerce in the colonial port accurately mirrored the mercantile system of London's port community. The majority of the weights have either legible guild stamps or what appear to be the remnants of indistinguishable marks which may be the remnants of cipher marks. The fact that pail weights with the angel figure were recovered as part of the assemblage is highly significant. The presence of these weights and the Archangel figure indicate an attempt by colonial authorities and merchants to adhere to the English metrological standard. Because most goods and merchandise that were shipped to the colony were packaged according to this standard, adherence to the legal standard appears to have been commonplace and the people of Port Royal may not have been living in what was, according to oral tradition, the "wickedest city in Christendom."

For over a thousand years in Western society, the iconography of the cardinal virtues, has served as the marker of social change and variance of value systems, both between and across ethnic and cultural boundaries. Adopting the strong, positive aspects of the Archangel Michael figure iconography was a wise tactical maneuver by both the Plumbers of London and the monarchy as a means of counteracting the long-standing tradition of regionally differentiated metrological standards. Foreign weightmakers brought not only a different set of technologies to the weight making trade in England - they also brought new influences in the form of mental templates to a country and a guild system that were in the midst of economic and social change. The melange of symbolic meanings entwined in the Archangel Michael figure are multi cultural and almost ageless in our history. Only an interdisciplinary research approach seems appropriate to unravel the host of meanings represented in the Angel-with-wings cipher.

The facts that great care was taken to adjust old weights in order to bring them to an acceptable standard, is a significant point. Interestingly, many of the methods used for adjusting the mass of weights had changed very little since the earliest recorded nested weights used throughout the Roman Empire. Similar to many bronze nested weights in the Streeter and Monti Christi collections, many of the Port Royal nested weight had been filed, lathed or drilled on their bottom surfaces as a means of bringing them to a new standard.

While trying to account for the inclusion of seemingly unlikely weight standards such as 55, 9, 7 and 2 pound 4 ounce weights, it seemed logical to investigate assize legislation pre-dating the 1692 earthquake at Port Royal. Laws regarding the regulation of weights, and in particular, the regulation of both the weight and selling prices for important staples such as bread, followed the consensus of mercantile activity, rather than actually setting the standard. In other words, while the Act of 8 Anne of 1710, did solidify prices for wheat and for the making and selling of bread, the main thrust of the assize was to codified in the form of laws, the accepted general mercantile activities of Old and New World merchants. The archaeological evidence from the Port Royal weight assemblage would indicate that many of the standards codified in the 1710 Assize of Bread, were in common used in the pre-earthquake community of Port Royal, Jamaica.

## REFERENCES CITED


Arnold, J. Barto, and Robert Weddle,1978
*The Nautical Archaeology of Padre Island, the Shipwrecks of 1554.* Academic Press, New York City, New York.

Atton, Henry, and H. Hurst Holland, 1967
*The King's Custom Volume 1.* Augustus M. Kelley, Publisher, New York City, New York.

Brewer, Thomas,1853
On the Antiquity of Markings and Stamping Weights and Measures. *Journal of the British Archaeological Association* 7:309-322.

Bronner, Simon J. ,1992
The Idea of the Folk Artifact. In *American Material Culture and Folklife, a Prologue and Analogue*, edited by Simon J. Bronner, pp. 3-39. University of Michigan Press, Ann Arbor, Michigan.

Buttrey, T., and Clyde Hubbard, 1971
*A Guide Book of Mexican Coins, 1822 to Date, Second Edition.* Western Publishing Company, Racine, Wisconsin.

Buttrey, Theodore V.,1907
*A Guide Book of Mexican Decimal Coins, 1863-1963.* Whitman Publishing Company, Racine, Wisconsin.

Cheminant, Richard Le, 1979
Antique Lead Merchant Weights from the City of London. *The London Archaeologist* 3(11):289-291.

Coit, Stanton, 1974
*Neighborhood Guilds, an Instrument of Social Reform.* Arno Press, The New York Times Company, New York City, New York.

Connor, R. D., 1987
*The Weights and Measures of England.* Her Majesty's Stationery Office, Science Museum of London, London, England.

Corporation of the City of London, 1698
*An Act of Common Council for the Better Regulation of Weighing Goods and Merchandises at the King's Beam.* Samuel Roycroft, Printer, London, England.

Corporation of the City of London, 1870
*On the Abolition of the Troy Weights.* Queens Printer, London, England.

Crellin, J. K., and J. R. Scott, 1969
*Wellcome Institute for the History of Medicine,* Pharmaceutical History and its Sources in the Wellcome Collections; Part II, Drug Weighing in Britain, c.1700-1900. 13:51-67.

Dahmus, Joseph, 1984
*Dictionary of Medieval Civilization.* Macmillan Publishing Company, New York City, New York.

Danforth, Ellen Zak, 1954
*Nesting Weights, Einsarzgewichte and Piles a Godets.* Archon Books, Hamden, Connecticut.

Dunn, Richard S., 1972
*Sugar and Slaves, the Rise of the Planter Class in the English West Indies, 1624-1713.* W. W. Norton & Company, New York City, New York.

Farmer, David Hugh, 1987
*The Oxford Dictionary of Saints, Second Edition.* Oxford University Press, New York City, New York.

Founders Company, The, 1991
*The Worshipful Company of Founders.* Maurice Stevenson, Librarian, Institute of Trading Standards Administration, London, England.

Graham, J. T., 1979
*Weights and Measures and Their Marks.* Shire Publications, Number 44, London, England.

Guildhall Act, 1672
*Act for the Weighing of Goods at the King's Beam.* King's Printer, London, England.

Hallock, W., and Herbert Wade, 1906
*Outlines of the Evolution of Weights and Measures and the Metric System.* Macmillan and Company, New York City, New York.

Hamilton, D. L., 1976

*Conservation of Metal Objects from Underwater Sites: A Study in Methods.* Texas Memorial Museum and The Texas Antiquities Committee, Joint Publishers, Austin, Texas.

Hamilton, D. L., 1990

A Decade of Under Water Research at Port Royal. *INA Newsletter* 17(20):4-7.

Hamilton, D. L., 1991

*Conservation of Cultural Resources I Text.* Unpublished manuscript on file at Department of Anthropology, Texas A&M University, College Station, Texas.

Hill, John, 1695

*A Great Dealer Refined: Being a Useful Companion for All Traders.* H. Rhodes Company, London, England.

Houben, Gerard, 1978

*European Coin Weights for English Coins.* Hub Tonnaer Company, Amsterdam, Holland.

Jamaica, The Laws of, 1681-1712

*The Acts of Common Law for Jamaica, Laws of Jamaica.* Jamaican Government Publication, Public Archives Office, Spanish Town, Jamaica.

Kisch, Bruno, 1975

*Scales and Weights.* Yale University Press, New Haven, Connecticut.

Kula, Witold, 1986

*Measures and Men.* Princeton University Press, Princeton, New Jersey.

Leeson, R. A., 1979

*Travelling Brothers, the Six Centuries' Road from Craft Fellowship to Trade Unionism.* George Allen and Unwin Limited, Ruskin House, London, England.

Manwayring, Henry, 1644

*The Sea-Mans Dictionary: An Exposition and Demonstration of All the Parts and Thinge Belonging to a Ship Together with an Explanation of All the Terms and Phrafes Ufed in the Practick of Navigation.* Printed by W. Godbid for Benjamin Hurlock, London, England.

Marx, Robert, 1967

*Excavation of the Sunken City of Port Royal, Jamaica, December 1965 to December 1966, a Preliminary Report*, Institute of Jamaica, Kingston.

Marx, Robert, 1968

*1968 Excavation of the Sunken City of Port Royal, Jamaica, December 1966 to December 1967, a Preliminary Report.* Jamaica National Trust , Kingston, Jamaica 1967.

Mencken, H. L., Editor, 1942
*A New Dictionary of Quotations on Historical Principles from Ancient and Modern Sources.* Alfred A. Knopf Company, New York City, New York.

Mercatante, Anthony S., 1988
*The Facts on File Encyclopedia of World Mythology and Legend.* Facts on File, Publishers, New York City, New York.

Moxon, Joseph, 1677
*Mechanick Exercises, or, the Doctrine of Handy-Works.* Joseph Moxon, Publisher, London, England.
- A&M holdings, Early English Book Series, Microform Department, film B, 3609

Nesmith, Robert I., 1955
*The Coinage of the First Mint of the Americas at Mexico City, 1536-1572.* The American Numismatic Society, Number 131, New York City, New York.

Nightingale, Pamela, 1985
The Evolution of Weight-Standards and the Creation of New Monetary and Commercial Links in Northern Europe from the Tenth Century to the Twelfth Century. *The Economic History Review*, 2nd series 38(2):192-209.

O'Boyle D. D. Patrick, 1967
*New Catholic Encyclopedia, Volume I.* McGraw-Hill Book Company, New York City, New York, for The Catholic University of America, Washington, D.C.

Ogilby, John, and William Morgan, 1677
*London Survey'd: or . An Explanation of The Large Map of London, Giving a Particular Account of the Streets and Lines in the City of Liberties with Courts, Yards, and Alleys, Churches, Halls, and Houses of Note in Every Street and Lane.* Printed by His Majesty's Printers, London, England.

Parsloe, Guy, 1964
*Wardens' Accounts of the Worshipful Company of Founders of the City of London, 1497-1681.* Oxford University Press for the Worshipful Company of Founders of the City of London, New York City, New York.

Pasley, M., 1834
*Observations on the Expediency and Actuality of Simplifying and Improving the Measures, Weights and Money Used in this Country, Without Materially Altering the Present Standards.* W. Burrill, Publisher, London, England.

Pawson, M., and D. Buisserret, 1975
*Port Royal, Jamaica.* Oxford University Press, Oxford, England.

Peterfreund, Stuart,1989
Blake on Charters, Weights, and Measures as Forms of Social Controls. *Studies in the Literary Imagination* 22(1):37-59.

Powel, John, 1671
*The Assize of Bread with Sundry Goods and Needful Ordinances for Bakers, Brewers, Inholders, Victuallers, Vintners and Butchers and Other Assizes in Weights and Measures.* Andrew Cook, Printer, London, England.

Roberts, Lewes, 1638
*The Merchants Mapp of Commerce Wherin the Universal Manner and Matter of Trade Is Compendiously Handled.* Ralph Mabb, London, England.

Royal Historical Society, 1962
*Transactions of the Royal Historical Society*, Fifth Series, Volume 12. Butler and Tanner Company Limited, Publishers for the London Office of The Royal Historical Society, London, England.

Roycroft, Samuel, 1698
*An Act of Common Council for the Better Regylation of Weighing of Goods and Merchandizes at the King's Beam.* City of London, Publishers, London, England.

Sheppard, William, 1675
*Of the Office of the Clerk of the Market of Weights and Measures and the Laws of Provision for Man and Beasts for Bread, Wine, Beer and Meal.* Samuel Heyrick Publisher, London, England.

Smith, Toulmin, 1963
*English Gilds, the Original Ordances of More than One Hundred Early English Gilds Together with Ye Olde Usages of Ye Cite of Wynchestre the Ordinances of Worcester the Ordinances of the Mayor of Bristol and the Costomary of The Manor of Tettenhall-Regis from Manuscripts of the Fourteenth and Fifteenth Centuries.* Oxford University Press, The Early English Text Society, New York City, New York.

Sokeland, Herrman, 1900
On Ancient Desemers or Steelyards. In *Annual Report of the Board of Regents of the Smithsonian Institution Showing the Operations, Expenditures, and Conditions of the Institution for the Year Ending June 30, 1900*, pp. 551-564. Smithsonian Institution Press, Washington, D.C.

Stampa, Manuel Carrera, 1949
The Evolution of Weights and Measures in New Spain. *Hispanic American Historical Review* February:2-24.

Stock, Ronald, 1973
The Stamping of Coin Weights, Part 2. *Libra* 10(3):7-10.

Thurston, Herbert, and Donald Attwater, 1962
*Butler's Lives of the Saints, Complete Edition, Volume III*, July, August and September. P. J. Kennedy & Sons, Publishers, New York City, New York.

Unwin, George, 1908
*The Guilds and Companies of London. Lectures on Economic History in the University of Edinburgh.* Methuen and Company, London, England.

Upton, Dell, 1985
The Power of Things: Recent Studies in American Vernacular Architecture. In *Material Culture, A Research Guide*, edited by Thomas J. Schlereth, pp. 57-78. University Press of Kansas, Lawrence, Kansas.

Weiner, Philip P. (editor), 1973
*Dictionary of the History of Ideas, Studies of Selected Pivotal Ideas, Volume IV, Psychological Ideas in Antiquity.* Charles Scribner's Sons, New York City, New York.

Wright, S. J. (editor),1988
*Parish, Church and People, Local Studies in Lay Religion, 1350-1750.* Century Hutchinson Company Limited, London, England.

Zupko, R. E., 1977
*British Weights and Measures: History from Antiquity to the Seventeenth Century.* University of Wisconsin Press, Madison, Wisconsin.

**Supplementary Sources Consulted**


Adams, Jonathan, 1986
*Sea Venture*, The International Journal of Nautical Archaeology 14(4):275-299.

Alexander, J. H., 1867
*Universal Dictionary of Weights and Measures, Ancient and Modern: Reduced to the Standard of the United States of America.* Van Nostrand Co., New York City, New York.

Arbuthnot, John, 1698
*Tables of Ancient Coins, Weights and Measures.* J. Tinson Co., London, England.

Arbuthnot, John, 1727
*Tables of Ancient Coins, Weights and Measures, Explain'd and Exemplify'd in Several Dissertations.* J. Tinson Co., London, England.

Avila, Manuel, 1975
*Diccionario Manual de Pesas y Medidas Recopilacion de Mas de Mil Unidades, Nacionales y Extrajeras, Anliguas y Modern.* Systema Internacional de Unidades, Buenos Aires.

Balasa, Bela, 1964
*Trade Prospects for Developing Countries.* Richard D. Irwin Co., Homewood, Illinois.

Berriman, A. E., 1969
*Historical Metrology: A New Analysis of the Archaeological and Historical Evidence Relating to Weights and Measures.* Greenwood Press, New York City, New York.

Blackhall, G., 1699
*Rules for Assizing of Bread: Troy Weights, Avoirdupois Weights.* Joseph Ray, Dublin, Ireland.

Burns, Arthur R., 1927
*Money and Monetary Policy in Early Times.* A. A. Knopf Co., New York City, New York.

Burton, William K., 1972
*Measuring Systems and Standards Organization.* American National Standards Institute, New York City, New York.

Clarke, F. W., 1882
*Weights, Measures and Money of All Nations.* A. Appleton and Company, New York City, New York.

Claypole, W., and D. J. Buisseret, 1972
Trade-Patterns in Early English Jamaica. *Journal of Caribbean History* 3:4-89

Considine, D. M., 1949
*Industrial Weighing*. Reinhold Publishing, New York City, New York.

Corporation of the City of London, 1672
*An Act of the Weighing of Goods at the King's Beam.* Andrew Clark, Printer for The City of London, London, England.

Evelyn, Sir George S., 1798
*An Account of Some Endeavours to Ascertain a Standard of Weights and Measures.* Phil. Trans. Roy. Society, London, England.

Everington, John, 1694
*The Gauger and Measures's Company.* James Lightbody Publisher, London, England.

Griffenhagen, George, 1956
*Pharmacy Museum, Tools of the Apothecary.* University of Wisconsin Press, Madison, Wisconsin.

Guerra, Francisco, 1960
Weights and Measures in Pre-Columbian America. *Journal of the History of Medicine and Allied Sciences* 28:535-537.

Hughson, Shirley Carter, 1984
*The Carolina Pirates and Colonial Commerce, 1670-1740.* Johns Hopkins Press, Baltimore, Maryland.

Huntar, Alexander, 1624
*A Treatise of Weights, Mets and Measures of Scotland.* Theatrum Orbis Terrarum, Amsterdam, Holland.

Ingalls, Walter Renton, 1865
*Systems of Weights and Measures.* The American Institute of Weights and Measures, New York City, New York.

Johns, Richard Anthony, 1988
*Colonial Trade and International Exchange*. Pinter Publishers, New York City, New York.

Kelemen, Pal, 1943
*Medieval American Art, a Survey in Two Volumes.* Macmillan Publishers, New York City, New York.

Kelemen, Pal, 1956
*Medieval American Art, Masterpieces of the New World.* Macmillan Publishers, New York City, New York.

Kelly, P., 1834
*Metrology, or an Exposition of Weights and Measures, Chiefly Those of Great Britain and France: Comprising Tables and Comparisons.* Publisher Unknown, London, England.

Kingdon, John Abernathy, 1905
*The Strife of the Scales, an Attempt to Explain How the King's Weigh House and Beams Within the City of London Came into the Charge of the Worshipful Company of Grocers.* Rixon and Arnold, London, England.

Klein, H. Arthur, 1974
*The World of Measurements; Masterpieces, Mysteries and Muddles of Metrology.* Simon and Schuster Publishers, New York City, New York.

Lang, Mabel L., 1964
*Weights and Measures and Tokens.* American School of Classical Studies at Athens, Princeton, New Jersey.

Lipson, E., 1948
*The Economic History of England, Volume II.* Adam and Charles Clock, London, England.

Long, Edward, 1970
*The History of Jamaica, Volume III.* Frank Cass Company, London, England.

Massey, J. Earl, 1968
*America's Money, the Story of Our Coins and Currency.* Thomas Y. Crowell Company, New York City, New York.

McCusker, John J., 1973
Weights and Measures in the Colonial Sugar Trade: The Gallon and the Pound and Their International Equivalents. *William And Mary Quarterly, 3rd series* 30 (4):599-624.

Moody, Ernest A., 1952
*The Medieval Science of Weights.* University of Wisconsin Press, Madison, Wisconsin.

Munoz, Miguel L., 1975
*Historia Numismatica del Estado de Mexico, Biblioteca Enciclopedia del Estado de Mexico.* Mexico City, Mexico.

Nicholson, Edward, 1901
*The Story of Our Weights and Measures.* Nicholson Publishing, Liverpool, England.

Pauston, Alexis J. P., 1780
*Metrologie, ou Traite des Mesures, Poids et Monnies des Ancien Peupleset des Modernes.* Veuve Desaint, Paris, France.

Petrie, Flinders, 1934
*Measures and Weights.* Metheuen and Company, London, England.

Sanders, L., 1960
*A Short History of Weighing.* W. & H. Avery Company, Birmingham, England.

Schildhauer, J., 1985
*The Hansa: History and Culture.* Edition Leipzig, Leipzig, Germany.

Sheppard, T., 1975
*Money Scales and Weights.* Organization of International Numismatists, London, England.

United States Government, 1955
*Statistical Office of the United States World Weights and Measures: Handbook for Statisticians, Provisional Edition.* Agricultural Organization of the United Nations, New York City, New York.

Wingwood, Allan, 1986
Sea Venture. *The International Journal of Nautical Archaeology* 20:149-159.

Zahedieh, Nuala, 1986
Trade, Plunder and Development in Jamaica, 1655-1689. *The Economic History Review*, 2nd series 29(2):205-222.

Zahedieh, Nuala, 1986
The Merchants of Port Royal, Jamaica, and the Spanish Contraband Trade, 1655-1692. *The William and Mary Quarterly*, 3rd series 43(4):570-593.

Zupko, R. E. 1985
*Dictionary of Weights and Measures for the British Isles.* University of Wisconsin Press, Madison, Wisconsin.

Zahedieh, Nuala, 1986
*A Dictionary of English Weights and Measures: From Anglo-Saxon Times to the Nineteenth Century.* University of Wisconsin Press, Madison, Wisconsin.

Zahedieh, Nuala, 1987
*The Weights and Measures of England.* University of Wisconsin Press, Madison, Wisconsin.

Zahedieh, Nuala, 1990
*Revolution in Measurement: Western European Weights and Measures since the Age of Science.* American Philosophy Society, Volume 186, Philadelphia, Pennsylvania.

**Appendix Tables**

**Catalog of Weights**

## APPENDIX A

## MARKS AND TERMS

| Mark | Description | Comments |
|---|---|---|
|  | Archangel Michael figure used by the Plumbers Company of London<br>- possibly first used by Msrs. de Grave and Company | - no marks of this style have been recovered from weights at Port Royal |
|  | Archangel Michael figure as found on two dome (pail) weights from Port Royal. | The angel figure is often shown in a full skirt or in a pair of hose. |
|  | Daggers were used as the insignia for the City of London and were adopted as a sign of quality craftsmanship. | Guild and craft houses outside of London established trade partnerships with London merchants in allowing them to use the insignia. |
|  | Ancient hand held dessemer or steelyard with lade pan. | The counterpoise of this style dessemer is hollow and a weight inscription was engraved along the length of the beam. |
|  | This is an equal arm scale or balance suspended from above. | The use of scales of this style date to circa 600 BC. |

| Mark | Description | Comments |
|---|---|---|
|  | The ewer or laver mark associated with the Founders Company of London, makers of bronze weights. | For a brief period of time, after 1655, the ewer stamp was reversed on bronze weights checked by the Founders. |
|  | The Troy mark was created by combining the initials "T" and R". | This mark is used universally today to denote the troy standard of measurement. |
| 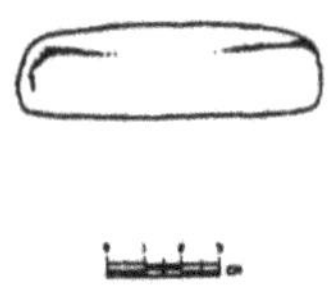 | Lead bun weights are usually regular in form but do not have raised upper or lower rims. | These weights are used on all types of scales. |
|  | Lead disc weights have raised rims similar in style to bronze disc weights. | These weights are commonly used on small to medium sized scales for commercial transactions. |
| 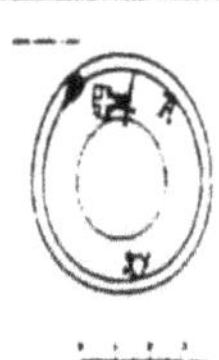 | Bronze disc weights usually have a raised rim on the top surface. The presence of a bottom rim is rare. | Bronze disc weights are diagnostically important since they were stamped with regis marks which identify a monarch, the city of origin and weight designations. |

| Mark | Description | Comments |
|---|---|---|
|  | Pail weights are a type of dome or tall weight made by pouring lead into an iron sleeve. | Weights such as these were commonly used on larger scales and their iron handles allowed easy transport and suspension from a beam arm. |
|  | Square weights were used similarly to pail weights. These weights could either suspended from a hook or placed on the pan of a scale. | Because of the protection that the large iron handle offers the top surface of the weight, all of the square weights from the site have excellent marks. |
|  | Cone weights were mad by forming a cone of copper into which lead was poured. These weights were used on all types of scales and a hook shaped handle extending from the top of the cone allowed them to be suspended from a beam. | PR90 897-9.1 is an unusual weight since it was filled with sand. Unlike other cone weights, this weight appears to have had a base which was pressure fit into the body of the weight. With this design, the weight was easily adjustable since sand could either be added of subtracted. |

## APPENDIX B

## WEIGHTS IN INVENTORIES

| NAME | VOL.& FOLIO | ITEMS |
|---|---|---|
| Barras, Anthony | V3.f.231-33 | half of a pr. of stilliards |
| Barre, Charles | V3.f.254-65 | 1 brass scale & pile of weights & a pair of stilliards |
| Bedow, John | V3.f347-348 | a pair of scales and a turtle shell case |
| Benning, Simon | V3.f.63-64 | 2 pr. of old scales & weights |
| Biggs, John | V3.f.221-2 | 2 pr. of stilliards |
| Bonham, Stephen Capt. | V3.f.379 | a pair of stillyards |
| Booker, Charles | V3.f.112-13 | a pr. of scales & stillyard |
| Bull, William | V3.f373-374 | a pair of old stillyards |
| Butler, George | V3.f.272-73 | 1 pr. of stilliards |
| Cheston, Edward | V3.f.209 | a pr. of stilliards & a pr. of scales |
| Clarke, Thomas | V3.f.349 | a pair of stillyards |
| Conner, Mathew | V3.f337-338 | two pair brass scales and 10 lbs. of lead, one pair of wooden scales |
| Coulson, Samuell | V3.f.246-47 | a beam scale & waites, 1 box with waites & scales & 1 silver scale |
| Cransborough, Nicholas | V3.f383-385 | 2 pair old stillyards, 1 pair of scales and weights |
| Davis, William | V3.f.297-98 | 1 beame with scales & weights, 2 pr. of small brass scales & weights & 2 small paires of scales & weights for money |
| Egelton, Henry | V3.f.48-49 | 2 pr. scales |
| Elkin, Isaac | V3.f.262-63 | 1 pr. of stilliard |

| NAME | VOL.& FOLIO | ITEMS |
|---|---|---|
| Faucett, Magdalen | V3.f.233-35 | one pr. stilliards |
| Fox, Edward | V3.f.280-83 | 1 paire of stilliards w. 28 |
| Freeman, Judith | V3.f.104-8 | 1 pr. of stilliards |
| Gardner, John | V3.f.265 | 1 pr.of stilliards |
| Guepin, John | V3.f.241-43 | 2 pr. brass scales & waites |
| Hayes, William | V3.f.291-96 | beame scales & weights & a pr. of small scales |
| Hazle, Edward | V3.f375-376 | 4lbs brass troy weights with beame and scales, 7.5 brass weights Haverdepois with beame and scales and 6 lbs lead - a P.R.merchant |
| Heard, Thomas | V3.f.239-40 | 2 pr. stilliards |
| Herring, Julines | V3.f350-352 | a pair of old brass scales and 4 lb. weight and one pair of stillyards |
| Hickes, Joseph | V3.f356-357 | one pair old stillyards and 49 lbs. weights |
| Hicks, Daniel | V3.f.249-52 | 2 small beams and scales with iron chains with 28 iron weights and 28,14,7,4,2,&1 - some of the Iron weights being broken |
| Ireland, John | V3.f386-387 | a box of scales and weights and 41.75 oz. silver in "Haverdpois" weight |
| Keeth, Alexander | V3.f.301 | one pair of scales and weights |
| Lamburne, Samuell | V3.f.262-62 | a beame with weights and scales |
| Launce, William | V3.f355-356 | one pair old stillyards and pea (?) |
| Mann, Samuell | V3.f.390 | a parcel of old scales & two pair small scales and beames |
| Marshall, John | V3.f323-324 | one pair of stillyards |

| NAME | VOL.& FOLIO | ITEMS |
|---|---|---|
| Modyford, Charles Sir | V3.f.55 | 3 pr. of stilliards, 2 small boxes Qts, 3 small p.scales & 3 setts of small Troy weights |
| Modyford, Charles Sir | V3.f.78-102 | 1 pr. of weights & weights for silver and small old stilliards, a beam & scale, 1 pr. of scales beame & leaden weights - 1 pr. of scales, beame, five hundred & twenty-eight pounds Iron weights & small weights about twenty-one pounds |
| Moone, Thomas | V3.f299-300 | a small beame, brass scales and small weights |
| Morant, John | V3.f.11-12 | a pr. of small stillyard |
| Morgan, Sir Henry | V3.f.258-61 | a box of Troy weights,a pair of stilliards |
| Municatt, Peter | V3.f.205 | 1 brass pann & 1 pr. small stilyards |
| Narnes, John | V3.f361-317 | 1 pair of scales and weights & a beame and scales |
| Neal, William | V3.f329-330 | lead & brass weights, an old iron beame and scales anda brass pair of scales with a beame |
| Newberry, Jobe | V3.f.335 | one pair of stillyards, two pairs of scales and some weights |
| Norris, Robert | V3.f.278-80 | one box of weights and scales & on the plantation, 1 old beame and scales & 2 old waites (28 & 14) |
| Norris, Thomas | V3.f.28-30 | 4 beame scales& a few waites. a small pr. of scale waites & a small pr. of scales |
| Perkins, William | V3.f.230-31 | a sorry pr. of old stilliards |

| NAME | VOL.& FOLIO | ITEMS |
|---|---|---|
| Phillpott, John | V3.f.285-90 | 6 scales, 3 beams, a parcell of scales, 100wt of brass weights, & 2 prs. of scales<br>- a known blacksmith |
| Pocock, Thomas | V3.f303-304 | 1 pair of scales and beame, & 2 pair of small old scales & weights |
| Pynck, Anthony | V3.f.353 | a pair of old stillyards |
| Rede, Henry | V3.f.203-4 | a box with weights & a pr. of stilliards |
| Richardson, Dorothy | V3.f.54 | 2 pr. of brass scales & weights |
| Roberts, Richard | V3.f312-314 | 1 pair of stillyards |
| Robinson, William | V3.f.26-27 | 2 pr. brass scales & weights |
| Roe, Benjamin | V3.f.273 | 1 small pr. of scales & weights |
| Salmon, John | V3.f354-355 | two pairs of stillyards |
| Scott, Robert | V3.f.15-16 | 8 pr. scales brass with some lead waites, 1 pr. scales |
| Scrivener, Peter | V3.f.275 | 1 paire of stilliards |
| Slade, William | V3.f314-315 | a pair of stillyards, a pair of old scales & 3 leaden weights |
| South, Thomas | V3.f.219-21 | 1 pr. of scales,1 pr. of stilliards & 1 pr. of scales & waites |
| Stanley, Robert | V3.f.17-18 | 2 brass scales |
| Thare, Samuel | V3.f.215-16 | a pr. of old brass scales |
| Thomas, Christopher | V3.f.265-67 | 1 pr. of brass scales, and he had lots of tobacco (incl. leaf) |
| Thompson, John | V3.f360-361 | one pair of stillyards |
| Tull,John | V3.f321-322 | a small meame - P.R. merchant |
| Unknown | V3.f.274-75 | two prs. of scales |
| Waite, Thomas | V3.f.238-39 | a wooden box with scales & waites |

| NAME | VOL.& FOLIO | ITEMS |
|---|---|---|
| Warne, Josia | V3.f.66-69 | a box of waites & a scale rule |
| Warner, Robert | V3.f.2-3 | a pr. of stilliards |
| Welling, Thomas | V3.f.245-6 | weights and scales, a pr. of stilliards & a pr. of brass scales - he may sell tobacco |
| Woosnam, Adam | V3.f380-381 | a pair old stillyards |

## INDEX C

## DE GRAVE AND COMPANY

| Company Name | Reference |
|---|---|
| Messrs De Grave and Company<br>- possibly brothers<br>- London circa 1750<br>- scales and weight makers<br>- used the Archangel Michael figure | - Le Cheminant<br>- 1979, p.190-1 |
| De Grave and Son<br>- London circa 1817<br>- makers of "coffin" shaped gold balances<br>- suspended brass construction | - Danforth<br>- 1986, entry 63<br>- unpublished manual for the management of the Edward Clark Streeter Collection of Weights and Measures |
| De Grave, Short and Fanner<br>- London circa 1850<br>- makers of apothecary scales and pharmaceutical weights | - Danforth<br>- 1986, entry 76 |
| Avery De Grave<br>- London, 1960<br>- maker and adjuster of grain scales and scientific scales | - Kisch - notes in a ledger of one of his file cards |

## APPENDIX D

## CATALOG: PORT ROYAL WEIGHT ASSEMBLAGE

CATALOG NUMBER: 1 ARTIFACT NUMBER: N.P.

EXCAVATION: Marx

WEIGHT TYPE: bun weight

MARKS: none

ORIGIN: European or locally made

ILLUSTRATIONS:

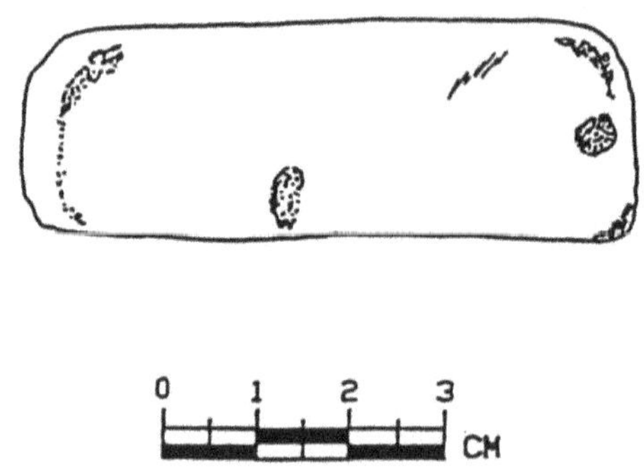


DESCRIPTION: A very flat circular weight with well-rounded edges -was fashioned from solid lead.

DIMENSIONS: HEIGHT (cm):1.90
DIAMETER/BASE (cm):7.40

WEIGHT IN KILOS: .91 WEIGHT IN POUNDS: 2 lb.

PROBABLE WEIGHT: 2 lb A. VARIANCE: 0.0%

PARALLELS: Several from the Port Royal collection.

MATERIALS: lead

COMMENTS: This weight could have been locally made. There are no marks to aid in identifying intended use or origin of the weight. With the rise in popularity of bronze weights because of the 1614 Charter of James I, 1614, lead weights fell into decreasing disuse.

CATALOG NUMBER: 2

ARTIFACT NUMBER: N.P.

EXCAVATION: Marx

WEIGHT TYPE: bun weight

MARKS: none

ORIGIN: Europe/local

ILLUSTRATIONS:

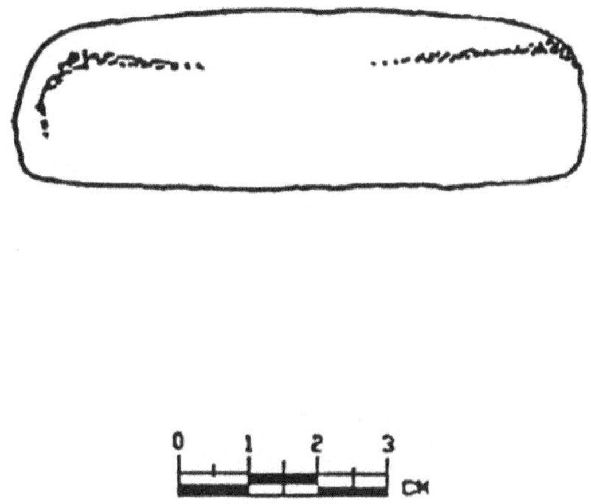


DESCRIPTION: The upper edge of the weight has large indentations and the bottom of the weight has deep score marks.

DIMENSIONS: HEIGHT (cm): 5.40
DIAMETER/TOP (cm): 6.40

WEIGHT IN KILOS: 1.02

WEIGHT IN POUNDS: 2.2486

PROBABLE WEIGHT: 2 lb 4 oz

VARIANCE : -.1%

PARALLELS: Same as entry #1

MATERIALS: lead

COMMENTS: At two points on the edge of this weight, the edges appear slightly flattened, as if the weight had been compressed. This weight has been identified as a bread weight, used to measure Sixpence bread by the Assize of Bread, dated 1710.

CATALOG NUMBER: 3

ARTIFACT NUMBER: N.P.

EXCAVATION: Marx

WEIGHT TYPE: bun/disc weight

MARKS: two possible marks

ORIGIN: Europe/local

ILLUSTRATIONS:

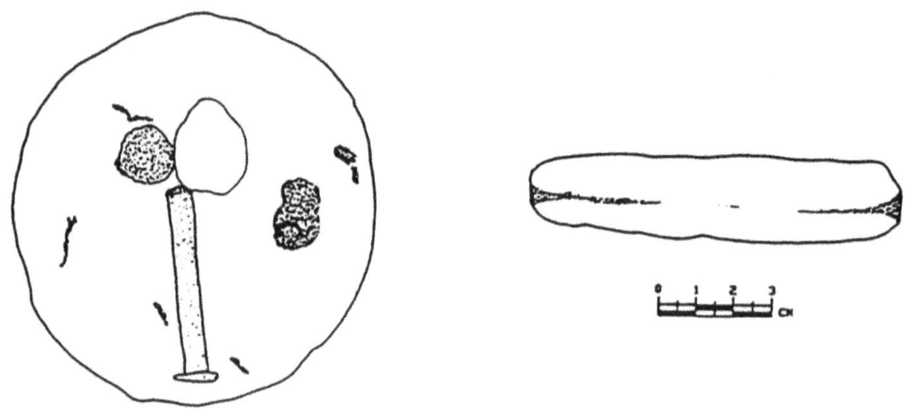

DESCRIPTION: This is a comparatively large, flat weight with a nail impression imbedded in the top surface. As with other weights from the site, nails had been used to alter the weight. Unfortunately, two possible marks on the top surface of the weight indistinguishable.

DIMENSIONS: HEIGHT (cm):1.9
DIAMETER/BASE (cm):10.2

WEIGHT IN KILOS: 1.82

WEIGHT IN POUNDS: 4.012

PROBABLE WEIGHT: 4 lb

VARIANCE: 0.0%

PARALLELS: The diameter and possible presence of marks make this weight unique on the Port Royal site.

MATERIALS: lead

COMMENTS: Artifact number PR89 857-9 has been adjusted in weight using large spikes.

CATALOG NUMBER: 4

ARTIFACT NUMBER: N.P.

EXCAVATION: Marx

WEIGHT TYPE: bun weight

MARKS: none

ORIGIN: Europe/local

ILLUSTRATIONS:

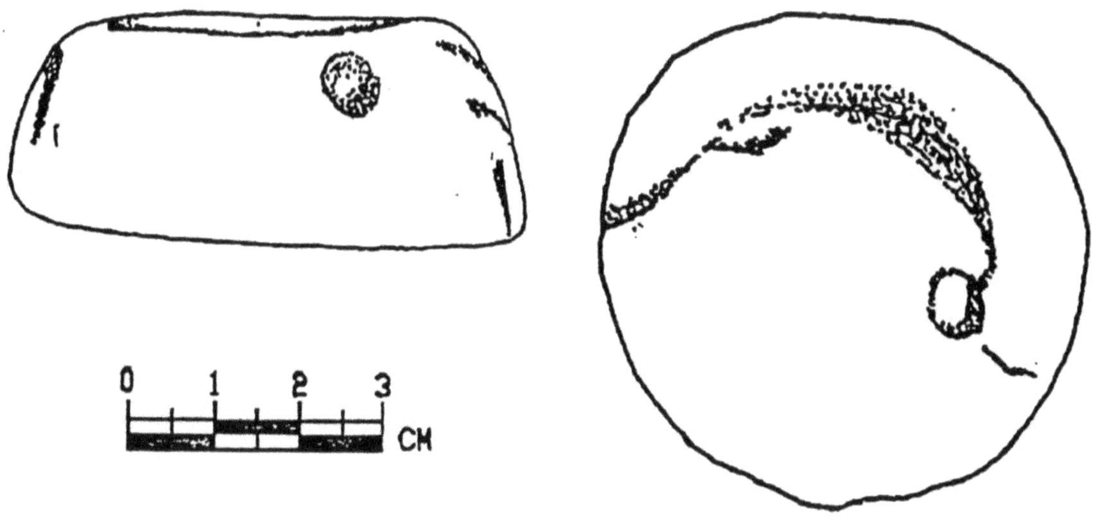


DESCRIPTION: As is typical with many bun weights, this weight has an irregular circular form and no distinguishable marks.

DIMENSIONS: HEIGHT (cm): 2.54
DIAMETER/BASE (cm): 5.35
DIAMETER/TOP (cm): 6.30

WEIGHT IN KILOS: 1.02

WEIGHT IN POUNDS: 2.29

PROBABLE WEIGHT: 2 lb 4 oz

VARIANCE: +1.8%

PARALLELS: same as entry 1

MATERIALS: lead

COMMENTS: bread weight

CATALOG NUMBER: 5 ARTIFACT NUMBER: N.P.

EXCAVATION: Marx

WEIGHT TYPE: bun weight

MARKS: none

ORIGIN: Europe/local

ILLUSTRATIONS:

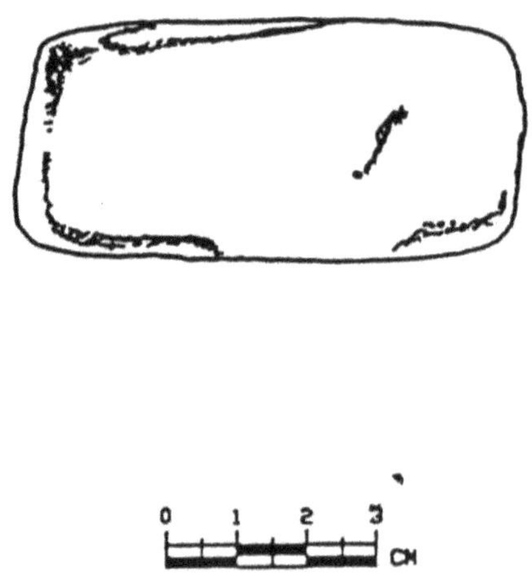


DESCRIPTION: The weight has an irregular, circular form with heavily worn edges. The surface of the weight is lightly pitted, and it is probable that the weight was made by casting lead in a sand form.

DIMENSIONS: HEIGHT (cm): 1.91
DIAMETER/BASE (cm): 5.35

WEIGHT IN KILOS: .91 WEIGHT IN POUNDS: 2.006 lb

PROBABLE WEIGHT: 2 lb A. VARIANCE: +.3%

PARALLELS: same as entry 1

MATERIALS: lead

COMMENTS: This weight is very similar to the rest of the bun weights from the site.

CATALOG NUMBER: 6 ARTIFACT NUMBER: CJ-93

EXCAVATION: Marx

WEIGHT TYPE: bun weight

MARKS: none

ORIGIN: Europe/local

ILLUSTRATIONS:

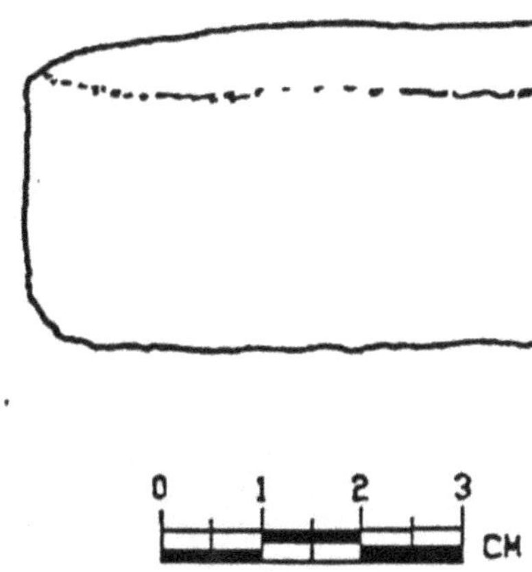


Side View

DESCRIPTION: This weight is large and circular in form with rounded edges on both the top and bottom surfaces. Both surfaces of the weight appear to be heavily abraded, possibly use ware.

DIMENSIONS: HEIGHT (cm): 1.60
DIAMETER/BASE (cm): 7.62

WEIGHT IN KILOS: .91 WEIGHT IN POUNDS: 2.006

PROBABLE WEIGHT: 2 lb A. VARIANCE: +.3%

PARALLELS: same as entry 1

MATERIALS: lead

CATALOG NUMBER: 7

ARTIFACT NUMBER: N.P.

EXCAVATION: Marx

WEIGHT TYPE: bun weight

MARKS: none

ORIGIN: Europe/local

ILLUSTRATIONS:

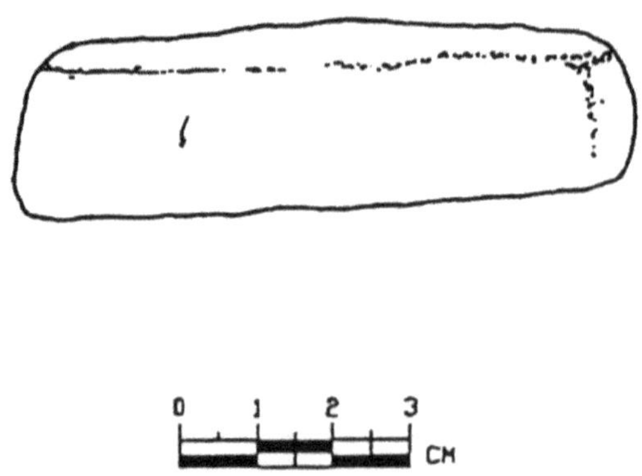


Side View

DESCRIPTION: The weight is small and circular in shape and was probably used as a pan weight for balance beams and scales.

DIMENSIONS: HEIGHT (cm): 1.27
DIAMETER/BASE (cm): 2.9

WEIGHT IN KILOS: 1.02

WEIGHT IN POUNDS: 2.25

PROBABLE WEIGHT: 2 lb 4 oz

VARIANCE: -.06%

PARALLELS: same as entry 1

MATERIALS: lead

COMMENTS: Bread weight.

CATALOG NUMBER: 8 ARTIFACT NUMBER: AJ-101

EXCAVATION: Marx

WEIGHT TYPE: bun weight

MARKS: none

ORIGIN: Europe

ILLUSTRATIONS:

Side View

DESCRIPTION: This weighs has an irregular, circular shape. It's flat appearance is similar to CJ-93.

DIMENSIONS: HEIGHT (cm): 1.60
DIAMETER/BASE (cm): 7.62

WEIGHT IN KILOS: .912 WEIGHT IN POUNDS: 2.010

PROBABLE WEIGHT: 2 lb A. VARIANCE: +.5%

PARALLELS: CJ-93

MATERIALS: lead

COMMENTS: The upper edge of the weight appears to have been slightly beveled.

CATALOG NUMBER: 9

ARTIFACT NUMBER: N.P.

EXCAVATION: Marx

WEIGHT TYPE: bun weight

MARKS: none

ORIGIN: Europe/local

ILLUSTRATIONS:

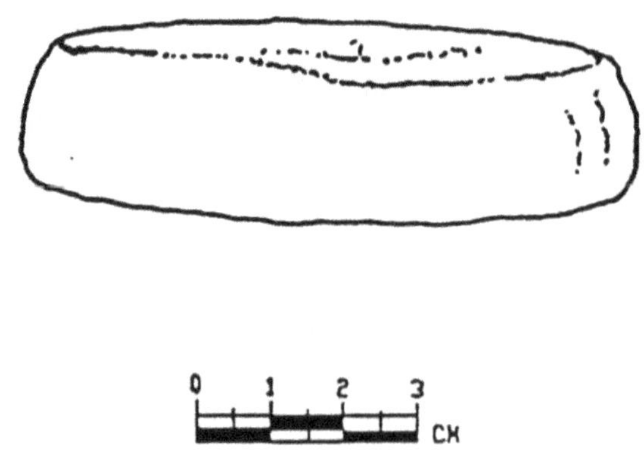


Side View

DESCRIPTION: This weight is small and nearly perfectly circular in shape with very little rounding of the edges and very few marks to indicate use-wear.

DIMENSIONS: HEIGHT (cm): 1.27
DIAMETER/BASE (cm): 2.90

WEIGHT IN KILOS: .45

WEIGHT IN POUNDS: .992

PROBABLE WEIGHT: 1 lb A.

VARIANCE: -.08%

PARALLELS: The same as several other weights, this one is small enough that it was probably used on a small balance beam or scale.

MATERIALS: lead

COMMENTS: none

CATALOG NUMBER: 10 ARTIFACT NUMBER: PR85 1024-2

EXCAVATION: Hamilton

WEIGHT TYPE: bun

MARKS: none

ORIGIN: Europe/local

ILLUSTRATIONS:

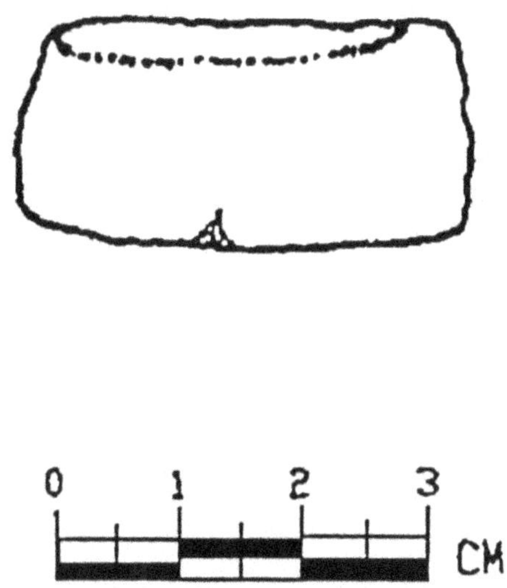


Side View

DESCRIPTION: The top surface of this weight is pitted and impurities in the texture of the lead indicate that this weight was crudely made. The weight is circular in shape and the top surface is slightly concave with a prominent rim.

DIMENSIONS: HEIGHT (cm): 1.90
DIAMETER/BASE (cm): 5

WEIGHT IN KILOS: .454 WEIGHT IN POUNDS: 1.000

PROBABLE WEIGHT: 1 lb A. VARIANCE: 0.0%

PARALLELS: Unlike most of the other bun weights, this one has a prominent raised rim. The top surface has the appearance of being hammered into shape and is very pitted.
MATERIALS: lead

COMMENTS: possibly locally made

CATALOG NUMBER: 11 ARTIFACT NUMBER: PR85 1026-6

EXCAVATION: Hamilton

WEIGHT TYPE: disc weight

MARKS: The weight has four large indentations on it's top surface which may have been from mark stamps.

ILLUSTRATION:

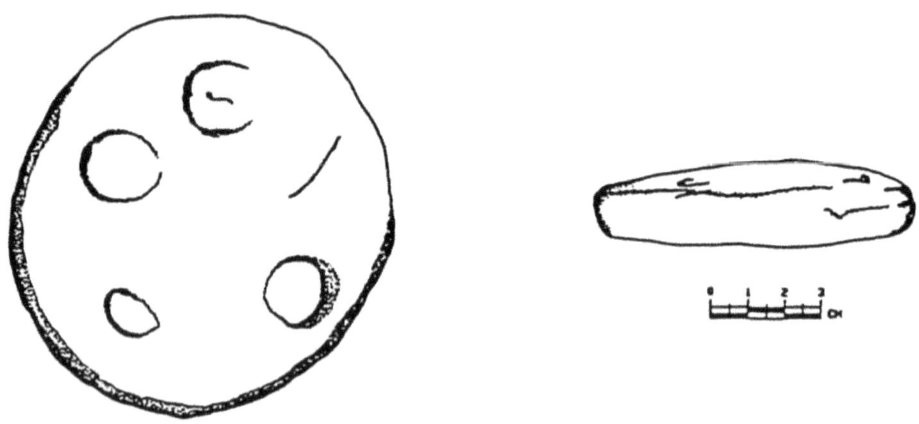

DESCRIPTION: The weight has mark indentations on it's top surface which are reminiscent of those found on dome weight PR85 1026-11.

DIMENSIONS: HEIGHT (cm): 2.55
DIAMETER/BASE (cm): 9.75

WEIGHT IN KILOS: 1.798 WEIGHT IN POUNDS: 3.9638

PROBABLE WEIGHT: 4 lb A. VARIANCE: -.1%

PARALLELS: This weight is similar in shape to CJ-93.

MATERIALS: lead

COMMENTS: The size and depth of the mark indentations on this weight indicates that this weight must have been founded and approved by the a guild foundry.

CATALOG NUMBER: 12 ARTIFACT NUMBER: PR85 1026-8

EXCAVATION: Hamilton

WEIGHT TYPE: bun weight

MARKS: - indication of two marks - neither distinguishable
- two parallel slash marks - possibly weight designation

ORIGIN: Europe/local

ILLUSTRATIONS:

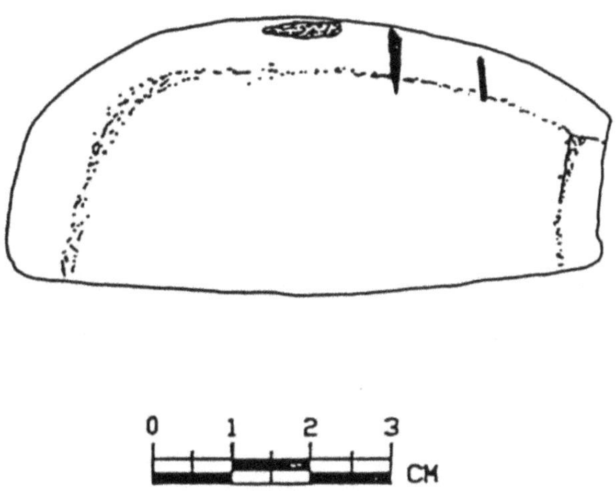


DESCRIPTION: The top surface of this weight is rounded and because of it's distinctive shape, the weight appears to have a manufactured look. A small section along one edge of the weight has been removed giving the edge a squared appearance.

DIMENSIONS: HEIGHT (cm): 1.90
DIAMETER/BASE (cm): 5.08

WEIGHT IN KILOS: .79 WEIGHT IN POUNDS: 1.7415

PROBABLE WEIGHT: originally 2lb A. VARIANCE: -13%

PARALLELS: none

MATERIALS: lead

COMMENTS: Apart from it's manufactured appearance, the weight has two parallel incised lines which may either be identification marks or one pound weight indicators. There are two indistinguishable marks on the top surface of the weight which may be assayers marks.

CATALOG NUMBER: 13 ARTIFACT NUMBER:PR85 1026-9

EXCAVATION: Hamilton

WEIGHT TYPE: bun weight

MARKS: - possible remnants of a mark on the top surface of the weight

ORIGIN: Europe

ILLUSTRATIONS:

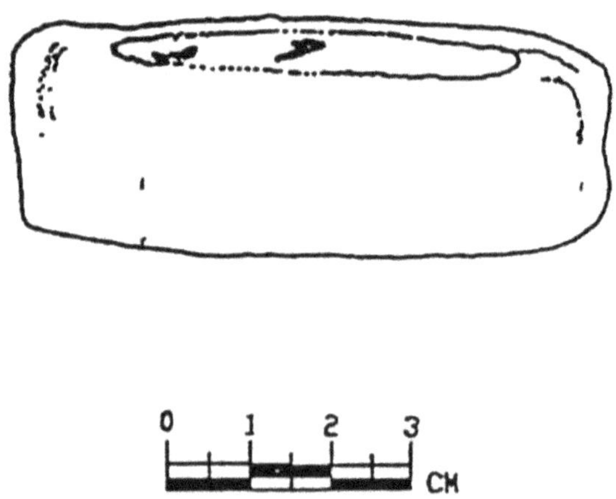


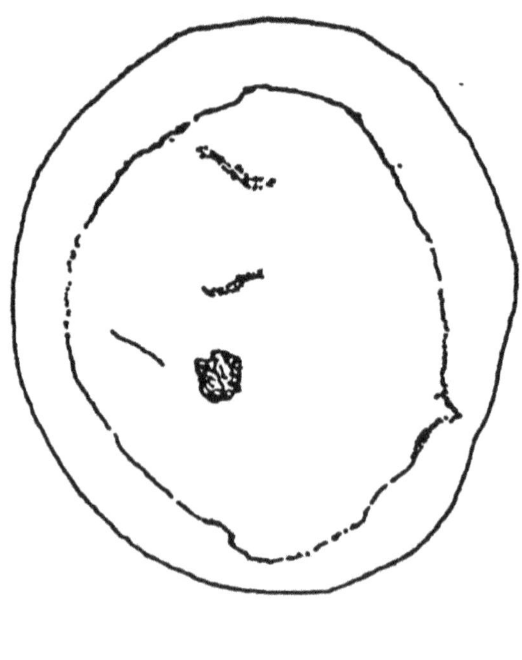

DESCRIPTION: This weight has an irregular, circular shape with a slightly concave top surface and a raised rim. Offset from the center of the top surface of the weight is a round indentation, probably left by a guild stamp or owners mark.

DIMENSIONS: HEIGHT (cm): 2.22
DIAMETER/BASE (cm): 6.40

WEIGHT IN KILOS: .22 WEIGHT IN POUNDS: .4850

PROBABLE WEIGHT: 8oz A. VARIANCE: -3%

PARALLELS: This weight is similar in shape and probable origin to PR85 1024.2.

MATERIALS: lead

COMMENTS: The weight was probably an officially sanctioned weight and probably was English in origin.

CATALOG NUMBER: 14 ARTIFACT NUMBER: PR85 1026-11.2

EXCAVATION: Hamilton

WEIGHT TYPE: bun

MARKS: Four circular indentations on the upper surface of the weight are probably for stamps and weight designations. Partial marks are visible in two of the four indentations.

ILLUSTRATION:

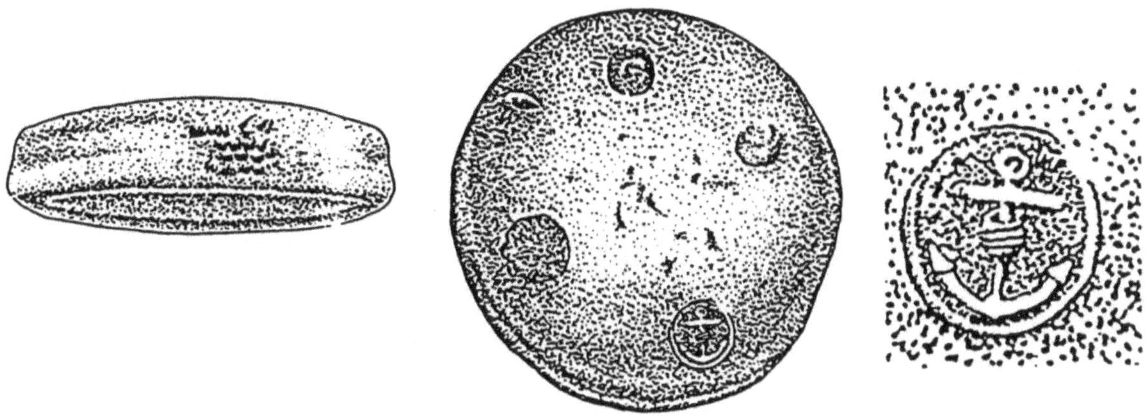

DESCRIPTION: The upper surface of this weight is badly scored and no marks are visible in the four indentations on the upper surface.

DIMENSIONS: HEIGHT (cm): 2.10
DIAMETER/BASE (cm): 98.00

WEIGHT IN KILOS: 1.798 WEIGHT IN POUNDS: 3.9638

PROBABLE WEIGHT: 4 lb A.2 VARIANCE: -1%

PARALLELS: Very similar to the PR85 1026-11.1 and PR85 1026-11.

MATERIALS: lead

COMMENTS: The upper surface of the weight appears to have been heavily abraded, making the marks indistinguishable.

CATALOG NUMBER: 15 ARTIFACT NUMBER:PR85 1026-11.3

EXCAVATION: Hamilton

WEIGHT TYPE: bun weight

MARKS: All four impressions on the top surface were created while marking the weight. A faint inscription is visible in only one indentation.

ORIGIN: Europe

ILLUSTRATIONS:

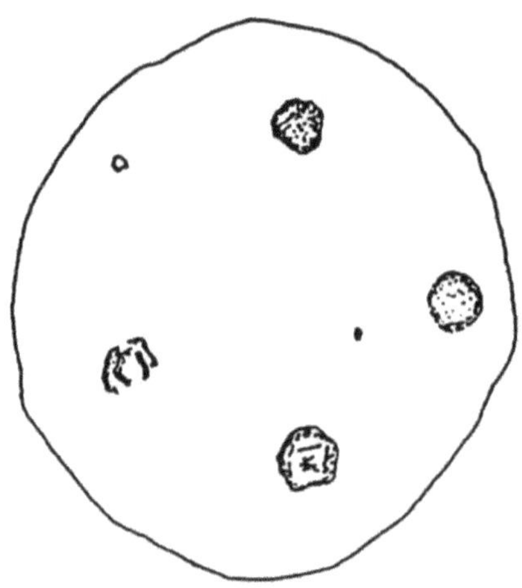

Four impressions of cipher marks

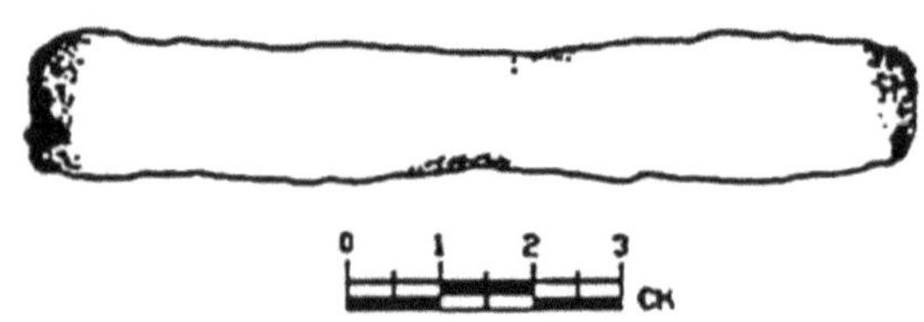


Side View

DESCRIPTION: Like PR85 1026-8, the weight has a rounded top surface and there is an indentation, possibly left by a makers mark or an assayers stamp. The mark is not legible.

DIMENSIONS: HEIGHT (cm): 2.15
DIAMETER/BASE (cm): 6.98

WEIGHT IN KILOS: .68 WEIGHT IN POUNDS: 1.499

PROBABLE WEIGHT: 1 lb 8 oz A. VARIANCE: 0.0%

PARALLELS: The shape of this weight is similar to that of PR85 1026-8.

MATERIALS: lead

COMMENTS: marks are illegible but this weight probable was sanctioned by the Plumbers Guild

CATALOG NUMBER: 16 ARTIFACT NUMBER: PR85 1026-11.4

EXCAVATION: Hamilton

WEIGHT TYPE: disc weight

MARKS: none

ILLUSTRATION:

Top View With Rim

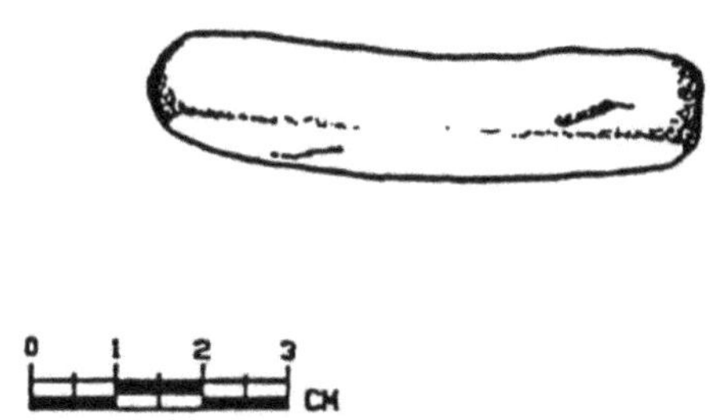


Side View

DESCRIPTION: This weight can be classified as either a bun or disc weight. Because it has a well defined circular shape and a raised rim, it has been included as a disc weight.

DIMENSIONS: HEIGHT (cm): 1.42
DIAMETER/BASE (cm): 6.65

WEIGHT IN KILOS: .21 WEIGHT IN POUNDS: .4629

PROBABLE WEIGHT: 8 oz A. VARIANCE: -7.4%

PARALLELS: Very similar to PR85 1026-9.

MATERIALS: lead

COMMENTS: The top surface of the weight has several lightly inscribed lines which have not been identified.

CATALOG NUMBER: 17

ARTIFACT NUMBER: PR85 1026-11.5

EXCAVATION: Hamilton

WEIGHT TYPE: bun weight

MARKS: none

ILLUSTRATION:

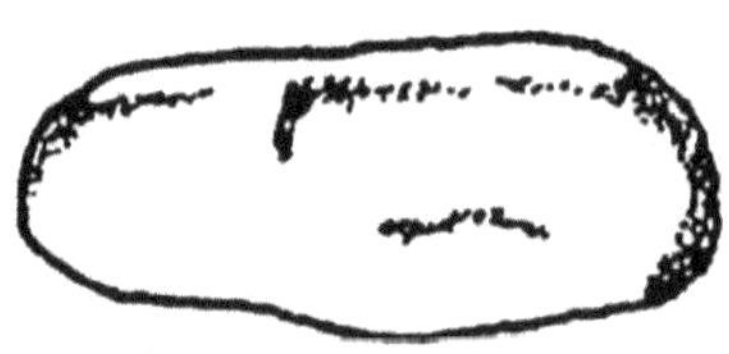

Side View

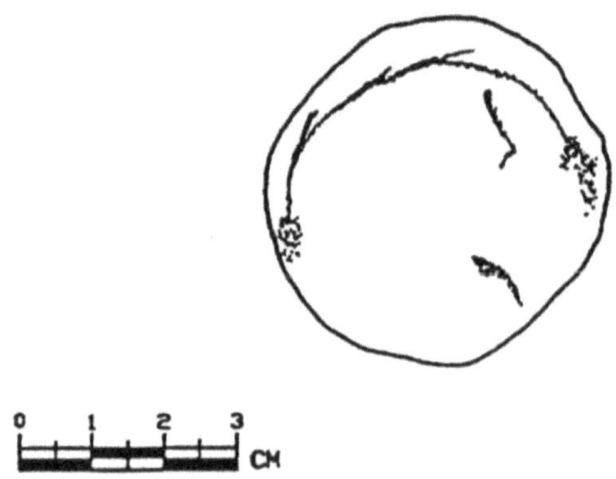


Top View Showing Portion Of Rim

DESCRIPTION: This weight is irregular in shape with a thickness that varies from 1.2 to 1.5 centimeters.

DIMENSIONS: HEIGHT (cm): 1.2 to 1.5
DIAMETER/BASE (cm): 4.652 (average)

WEIGHT IN KILOS: .12.

WEIGHT IN POUNDS: .2645

PROBABLE WEIGHT: 4 oz A.

VARIANCE: +5.8%

PARALLELS: This weight is similar to PR89 867-3.

MATERIALS: lead

COMMENTS: Weights such as this one were commonly used on small pan scales.

CATALOG NUMBER: 18 ARTIFACT NUMBER: PR85 1026-12

EXCAVATION: Hamilton

WEIGHT TYPE: disc weight

MARKS: none

ILLUSTRATION:

Top View Showing Rim

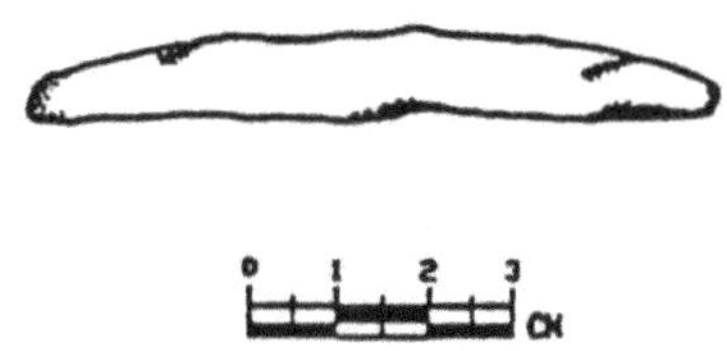


Side View

DESCRIPTION: This weight is very thin and irregular in shape. Located on the top surface of the weight is a very light indentation which may be the remnants of a mark. Because the surface had been heavily worn, the mark is not visible.

DIMENSIONS: HEIGHT (cm): .9
DIAMETER/BASE (cm): 9.4

WEIGHT IN KILOS: 1.798 WEIGHT IN POUNDS: 3.963

PROBABLE WEIGHT: 4 lb A. VARIANCE: -1.0%

PARALLELS: none

MATERIALS: lead

COMMENTS: This weight has a slightly raised rim and the edges of the weight are well rounded, indicating that the weight had been altered a great deal from it's original state.

CATALOG NUMBER: 19 ARTIFACT NUMBER: PR85 1034-2

EXCAVATION: Hamilton

WEIGHT TYPE: disc weight

MARKS: none

ILLUSTRATION:

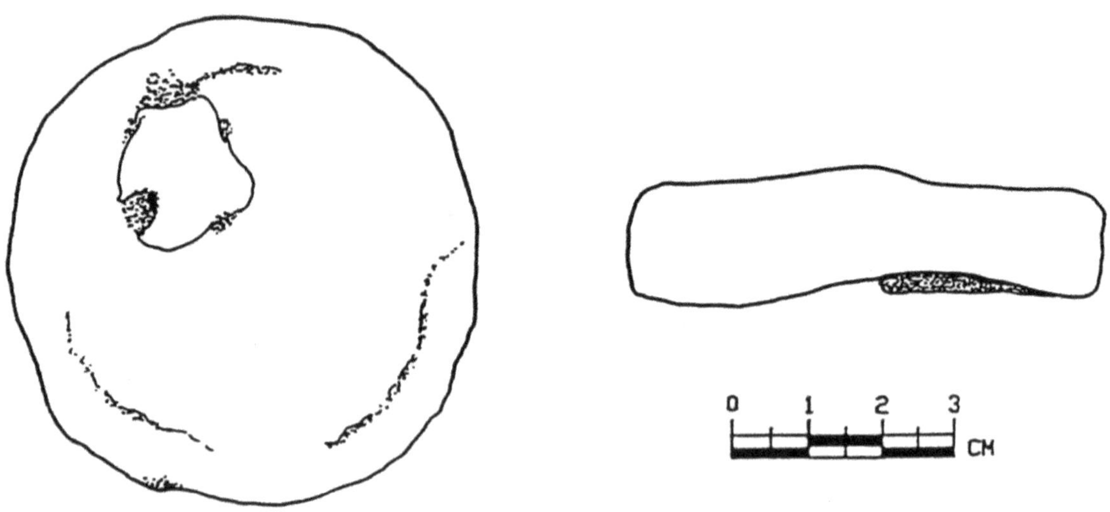


DESCRIPTION: This weight is very common in style except that the weight has had a lead patch added to it's bottom surface. The weight may have a slightly raised rim but this is speculation since the surface is heavily worn.

DIMENSIONS: HEIGHT (cm): 1.53
DIAMETER/BASE (cm): 6.420

WEIGHT IN KILOS: .448 WEIGHT IN POUNDS: .9876

PROBABLE WEIGHT: 1 lb A. VARIANCE: -1.3%

PARALLELS: This weight is very similar to PR 1034-7.

MATERIALS: lead

COMMENTS: No means of attaching the lead patch was noted in the records for the artifact. It appears from first hand inspection that the lead may have been melted into a partial void on what may have been the original top surface of the weight.

CATALOG NUMBER: 20 ARTIFACT NUMBER:PR85 1034-5

EXCAVATION: Hamilton

WEIGHT TYPE: bun weight

MARKS: none

ORIGIN: Europe

ILLUSTRATIONS:

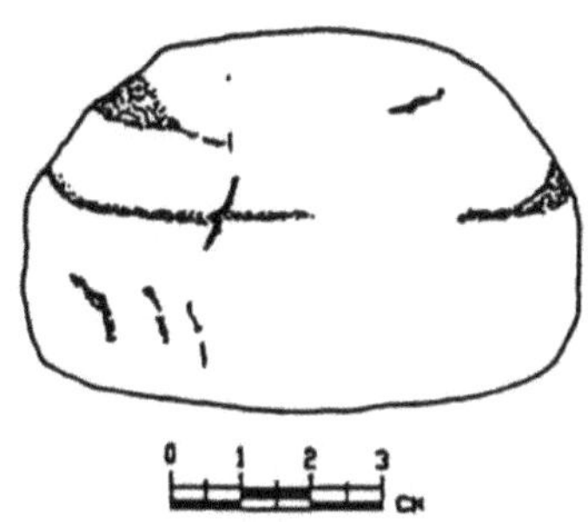


Side View

DESCRIPTION: This is a circular weight with a rounded top surface and flat, squared sides which appear to be sharply beveled up to the top surface of the weight. The surfaces of the weight are smoothly finished and like weight - surfaces of the weight are smooth and the weight appears to have a manufactured appearance like AJ-101.

DIMENSIONS: HEIGHT (cm): 4.27
DIAMETER/BASE (cm): 7.62

WEIGHT IN KILOS: 1.81 WEIGHT IN POUNDS: .3.99

PROBABLE WEIGHT: 4lb A. VARIANCE: -.3%

PARALLELS: This weight is similar to AJ-101,PR85 1026-8 and PR85 1026-11.3.

MATERIALS: lead

COMMENTS: This weight has a more manufactured appearance, based on its shape and smooth, polished finish.

CATALOG NUMBER: 21 ARTIFACT NUMBER: PR85 1034-5

EXCAVATION: Hamilton

WEIGHT TYPE: bun weight

MARKS: - obscured mark - probably a "regis" mark
- A - avoirdupois standard

ORIGIN: England

ILLUSTRATIONS:

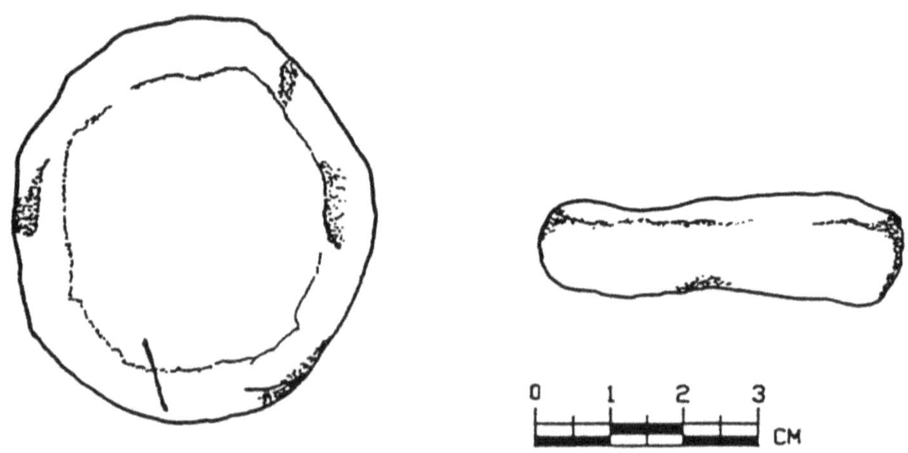


DESCRIPTION: The marks on this weight are partially obscured by extensive corrosion. The shape of an obscured mark is probably that of King Charles II since the Crown over "C" fits the outline of his mark.

DIMENSIONS: HEIGHT (cm): .64
DIAMETER/BASE (cm): 5.35

WEIGHT IN KILOS: .88 WEIGHT IN POUNDS: 1.94

PROBABLE WEIGHT: 2 lb A. VARIANCE: -3%

PARALLELS: The placement of the regis, dagger and "A" marks is unusual in that they face outward from the center of the weight and are positioned against the inner rim.

MATERIALS: lead

COMMENTS: This is a very typically styled disc weight except for the placement of the marks.

CATALOG NUMBER: 22 ARTIFACT NUMBER: PR85 1034-6

EXCAVATION: Hamilton

WEIGHT TYPE: disc weight

MARKS: - a lead adjuster had been nailed to the surface of this weight using a small square nail, which is visible on both surfaces of the weight.

ORIGIN: England

ILLUSTRATIONS:

DESCRIPTION: Portions of both the top and bottom rim of this weight are visible and while the iron nail fastener had disappeared, the lead adjuster is still in place on the surface of the weight.

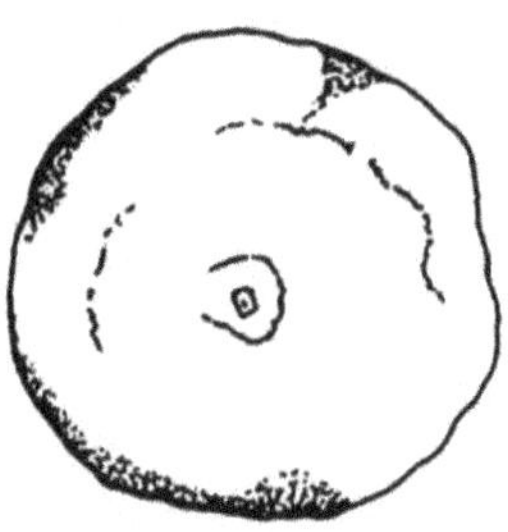

Top View

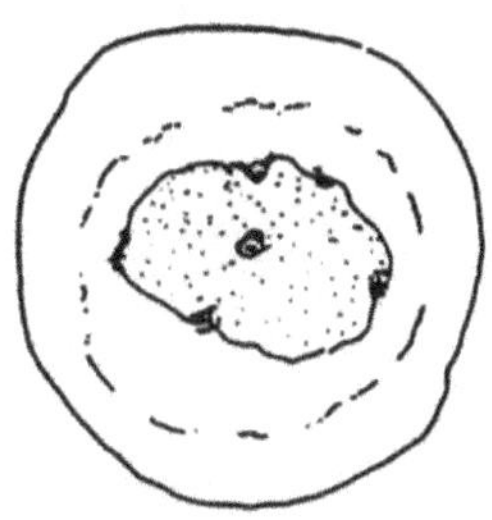

Bottom View

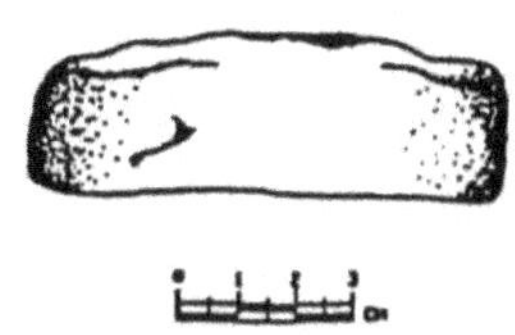

Side View

DIMENSIONS: HEIGHT (cm): .92
DIAMETER/TOP(cm): 6.52

WEIGHT IN KILOS:.194 WEIGHT IN POUNDS: .427

PROBABLE WEIGHT: 8 oz A. VARIANCE: -14.5%

PARALLELS: none

MATERIALS: lead

COMMENTS: This is a good example of a lead disc weight since both top and bottom rims are visible. Usually through use-wear, the bottom rim is badly worn. Possibly the addition of the adjuster gave added protection to the bottom of the weight so that much of the rim has been preserved.

CATALOG NUMBER: 23    ARTIFACT NUMBER: PR85 1034-7

EXCAVATION: Hamilton

WEIGHT TYPE: disc weight

MARKS: On the top surface of the weight are the faint outlines of blurred initials (possibly "AE").

ILLUSTRATION:

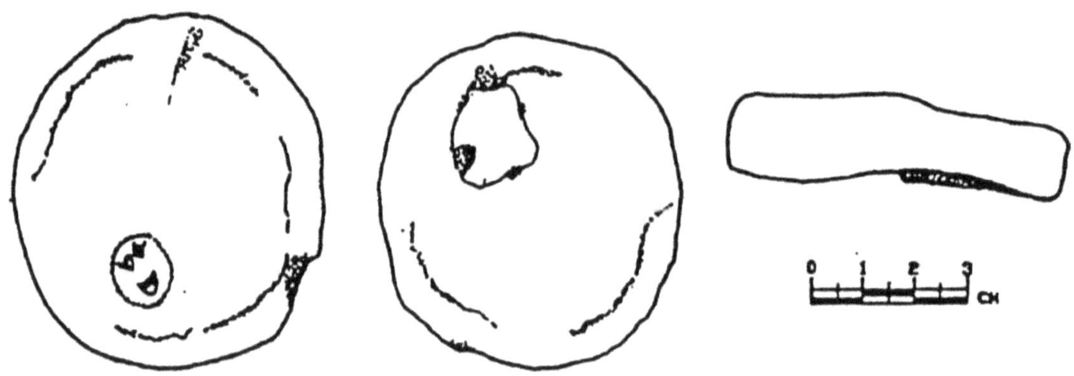


DESCRIPTION: This weight has a slightly raised rim on it's edges. The most distinguishing feature of the weight is circular mark indentation on the top surface with the initials "AE" inside a circle.

DIMENSIONS: HEIGHT (cm): 1.534
DIAMETER/BASE (cm): 6.436

WEIGHT IN KILOS: .448    WEIGHT IN POUNDS: .987

PROBABLE WEIGHT: 1 lb A.    VARIANCE: -1.23%

PARALLELS: This weight is similar to PR85 1024-2 which also has a slightly raised rim.

MATERIALS: lead

COMMENTS: none

CATALOG NUMBER: 24 ARTIFACT NUMBER: PR85 1034-10

EXCAVATION: Hamilton

WEIGHT TYPE: disc weight

MARKS: The weight has several, deeply inscribed lines on its bottom surface.

ILLUSTRATION:

Bottom Marks And Rim

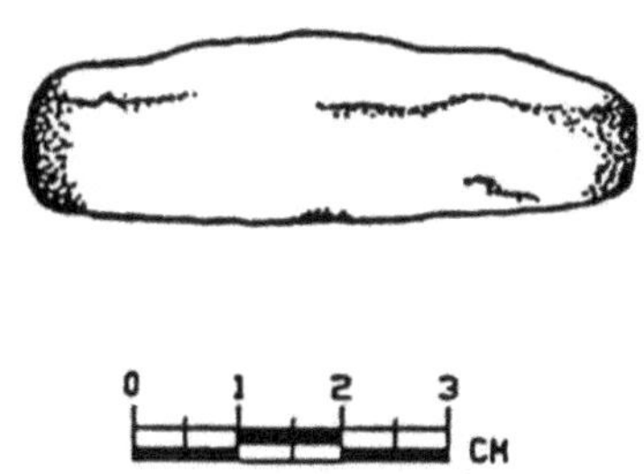


Side View

DESCRIPTION: The weight is circular in form and all surfaces are very rough and irregular. The edges surfaces of the weight are rounded and the top surface of the weight has a slightly raised rim.

DIMENSIONS: HEIGHT (cm): 1.14
DIAMETER/BASE (cm): 6.20

WEIGHT IN KILOS: .430 WEIGHT IN POUNDS: .947

PROBABLE WEIGHT: 1 lb A. VARIANCE: -5.2%

PARALLELS: The markings on this weight are very similar to those found on PR86 144-6.

MATERIALS: lead

COMMENTS: None of the inscribed lines have been identifies as specific markings and may be attributed to use-wear.

CATALOG NUMBER: 25 ARTIFACT NUMBER: PR85 1053-5

EXCAVATION: Hamilton

WEIGHT TYPE: bun weight

MARKS: - remnants of at least four nails probably used for adjusting the weight

ILLUSTRATION:

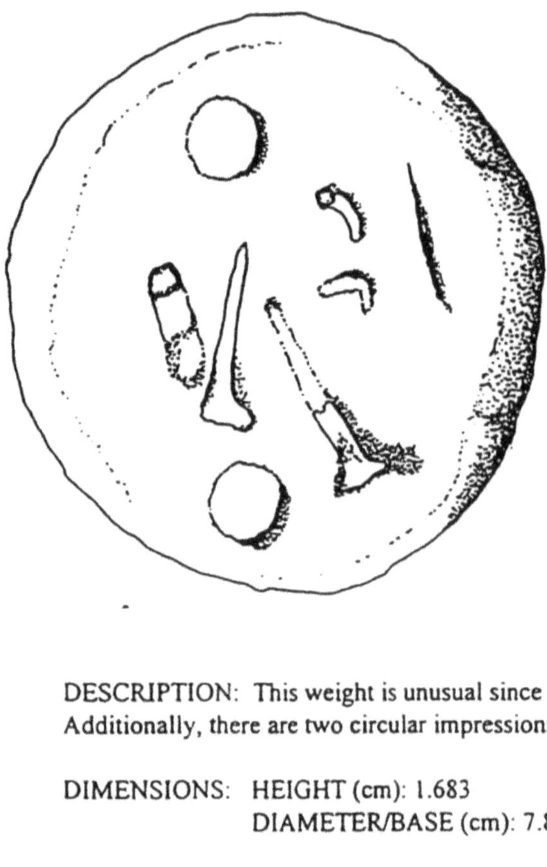

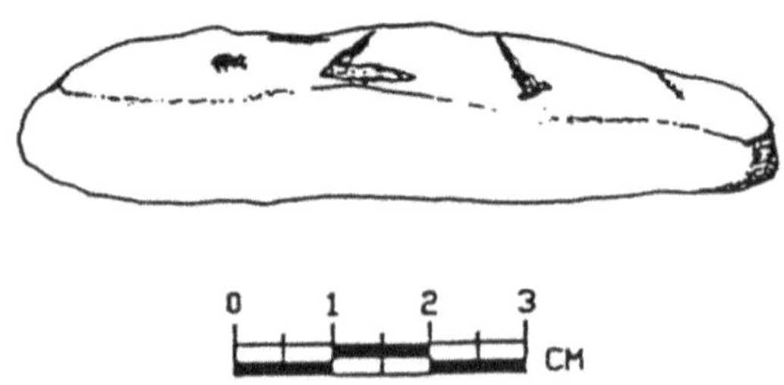


DESCRIPTION: This weight is unusual since it has markings from at least four nails on its top surface. Additionally, there are two circular impressions which probably contained guild stamps.

DIMENSIONS: HEIGHT (cm): 1.683
DIAMETER/BASE (cm): 7.845

WEIGHT IN KILOS: .8945 WEIGHT IN POUNDS: 1.972

PROBABLE WEIGHT: 2 lb A. VARIANCE: -1.7%

PARALLELS: This weight is very similar to one of Marx's no provenance bun weights which also contained the impression of a large nail.

MATERIALS: lead

COMMENTS: The nails used to adjust this weight are very large in comparison to the thickness of the weight.

CATALOG NUMBER: 26 ARTIFACT NUMBER: PR85 1043-11

EXCAVATION: Hamilton

WEIGHT TYPE: bun weight

MARKS: The weight has two worn mark indentations which are not legible

ILLUSTRATION:

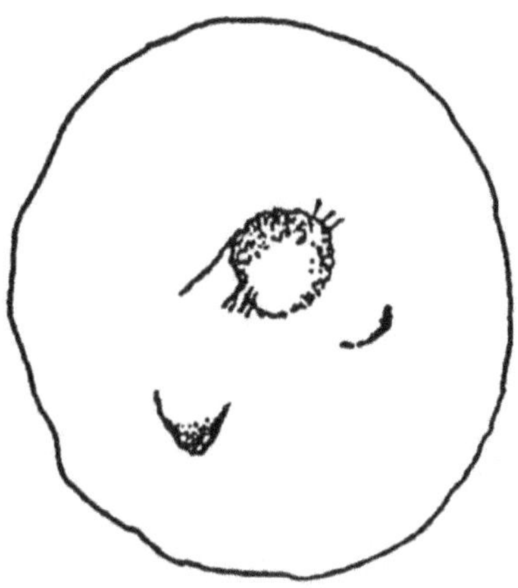

Top View With Marks

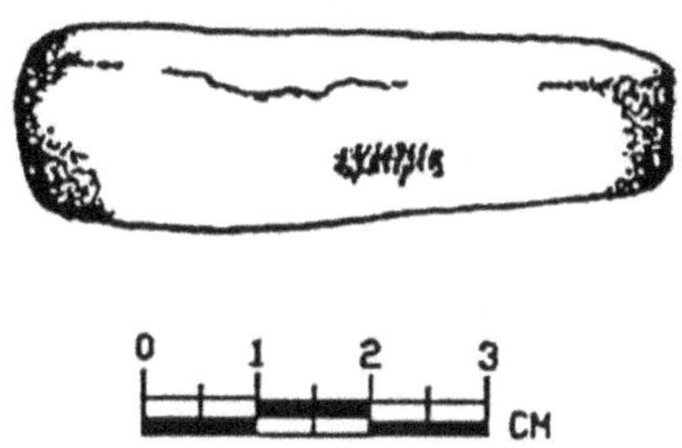


Side View

DESCRIPTION: The weight has an irregular circular shaped form with no distinguishing features beside the mark indentations.

DIMENSIONS: HEIGHT (cm): 1.64
DIAMETER/BASE (cm): 6.10

WEIGHT IN KILOS: .475 WEIGHT IN POUNDS: 1.047

PROBABLE WEIGHT: 1 lb A. VARIANCE: +5%

PARALLELS: This weight is similar in shape to several Port Royal weights.

MATERIALS: lead

COMMENTS: This is a very common weight form.

CATALOG NUMBER: 27

ARTIFACT NUMBER: PR86 144-6

EXCAVATION: Hamilton

WEIGHT TYPE: bun weight

MARKS: 5 incised lines

ILLUSTRATIONS:

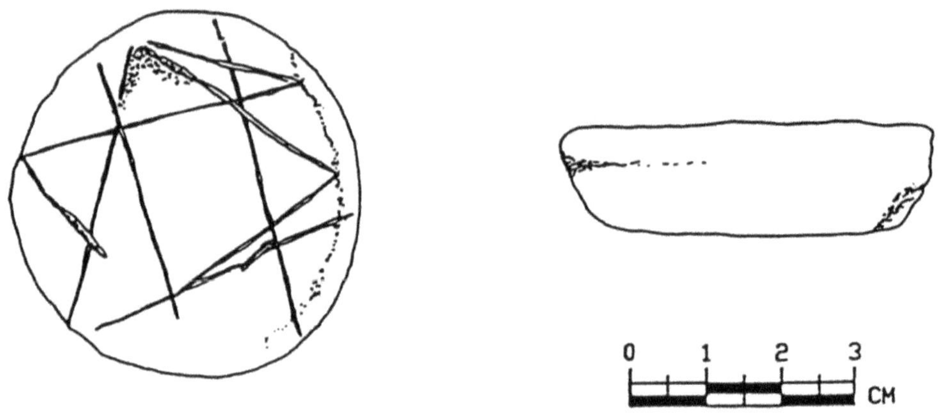


DESCRIPTION: The bottom surface (shown in the illustration) has several long incised lines, of which some are almost one millimeter in depth. The upper surface is smooth and the upper edge of the weight is rounded more than the bottom edge.

DIMENSIONS: HEIGHT (cm): 1.3847
DIAMETER/BASE (cm): 5.0438

WEIGHT IN KILOS: .2124

WEIGHT IN POUNDS: .468

PROBABLE WEIGHT: 8 oz A.

VARIANCE: -6.4%

PARALLELS: The inscribed lines on this weight are similar to those found on PR85 1034-10 and may have been a simple means of differentiating weights.

MATERIALS: Lead

COMMENTS: none

CATALOG NUMBER: 28 ARTIFACT NUMBER: PR89 743

EXCAVATION: Marx

WEIGHT TYPE: bun weight

MARKS: ♥ - heart in the center of the top surface - probably an owners mark

ORIGIN: unknown

ILLUSTRATIONS:

DESCRIPTION: The weight is irregular in shape and both the top and bottom surfaces are flat. In the center of the top surface is a deeply incised heart-shaped mark, as well as several parallel incised lines.

DIMENSIONS: HEIGHT (cm): 1.10
DIAMETER/BASE (cm): 4.75

WEIGHT IN KILOS: .224 WEIGHT IN POUNDS: .494

PROBABLE WEIGHT: 8 oz A. VARIANCE: -1.2%

PARALLELS: none

MATERIALS: lead

COMMENTS: In comparative terms, this is a very crudely made weight which was probably locally made.

CATALOG NUMBER: 29 ARTIFACT NUMBER: PR89 867-3

EXCAVATION: Hamilton

WEIGHT TYPE: bun weight

MARKS: Inscribed "S" on top and bottom surfaces

ILLUSTRATIONS:

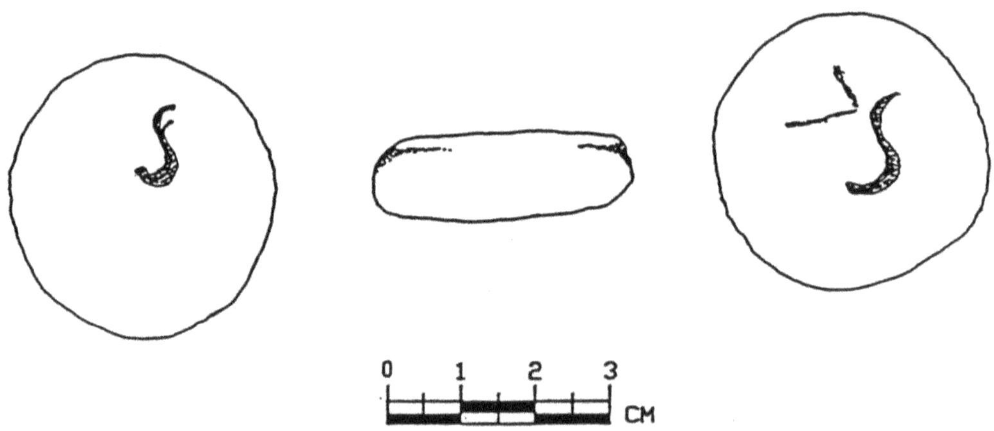


DESCRIPTION: This weight has a deeply incised "S" initial inscribed on both the top and bottom surfaces. One "S"appears to have been miss-struck which may explain an extra, shallower, top loop to the initial.

DIMENSIONS: HEIGHT (cm): 1.0595
DIAMETER/BASE (cm): 3.692

WEIGHT IN KILOS: .1104 WEIGHT IN POUNDS: .243

PROBABLE WEIGHT: 4 oz A. VARIANCE: -2.8%

PARALLELS:

MATERIALS: lead

COMMENTS: The "S" mark is probably an owners mark which to date, has not been identified.

CATALOG NUMBER: 30 ARTIFACT NUMBER: N.P.

EXCAVATION: Marx

WEIGHT TYPE: cylindrical weight

MARKS: none

ORIGIN: possibly Jamaica

ILLUSTRATIONS:

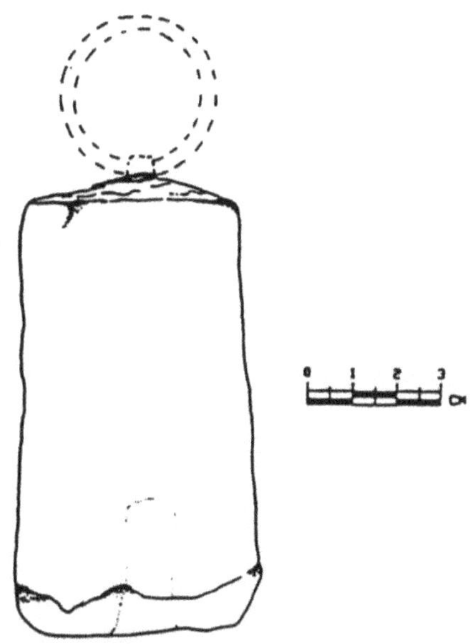

DESCRIPTION: This weight appears to have been made by hammering lead into a roughly cylindrical shape. An iron ring (missing) would have been attached to the body of the weight by a flanged, U-shaped bracket on the top surface. The hole in the base of the weight, which extends 3.5 cm. into the interior of the base was possibly used for adjusting the weight.

DIMENSIONS: HEIGHT (cm): 10.16
DIAMETER/TOP(cm): 5.72
DIAMETER/BASE(cm): 6.00

WEIGHT IN KILOS: 3.18 WEIGHT IN POUNDS: 7.01

PROBABLE WEIGHT: 7 lb A. VARIANCE: 0.0%

PARALLELS: none

MATERIALS: lead body - iron ring

COMMENTS: Weights such as this were commonly used on larger balance beams and scales.

CATALOG NUMBER: 31 ARTIFACT NUMBER: PR 98

EXCAVATION: Marx

WEIGHT TYPE: dome weight

MARKS: none

ORIGIN: unknown

ILLUSTRATIONS:

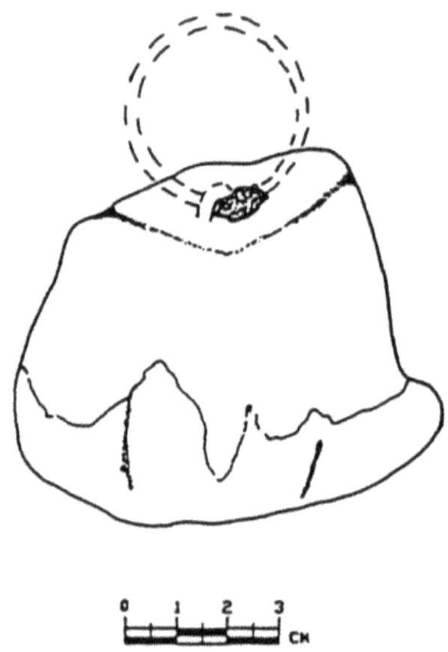


DESCRIPTION: The base of the weight shows signs of having been used extensively because the edges of the base of the weight have been pushed upwards, both accentuating the rim of the weight and making the weight appear to be compressed.

DIMENSIONS: HEIGHT (cm): 8.26
DIAMETER/TOP(cm): 5.82
DIAMETER/BASE(cm): 8.90

WEIGHT IN KILOS: 3.88 WEIGHT IN POUNDS: 8.55

PROBABLE WEIGHT: 9 lb A. VARIANCE: -5%

PARALLELS: This weight is similar to PR101.

MATERIALS: lead body - iron ring

COMMENTS: Weights of this style were used on scales and balance beams.

CATALOG NUMBER: 32

ARTIFACT NUMBER: PR 101

EXCAVATION: Marx

WEIGHT TYPE: dome weight

MARKS: 2 obscured marks

ORIGIN: unknown

ILLUSTRATIONS:

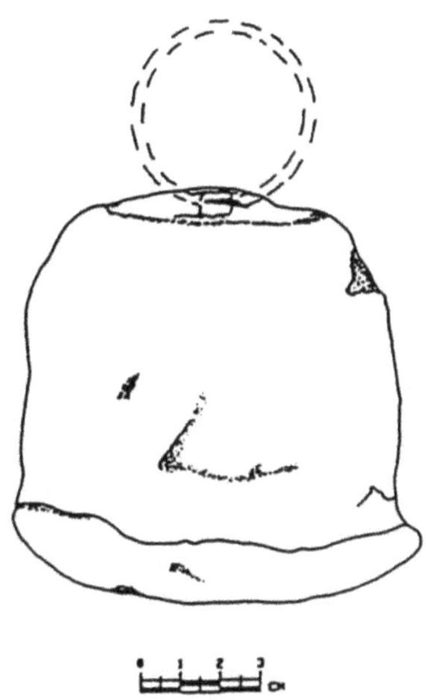

DESCRIPTION: This weight is a large bell weight which had an iron ring attachment for carrying the weight and loading it on a scale or balance beam. The weight has no visible marks.

DIMENSIONS: HEIGHT (cm): 9.398
DIAMETER/BASE (cm): 9.85

WEIGHT IN KILOS: 6.136

WEIGHT IN POUNDS: 13.53

PROBABLE WEIGHT: 14 lb A.

VARIANCE: -3.4%

PARALLELS: Similar in "squashed" appearance to pr 98.

MATERIALS: lead body - iron ring handle on top

COMMENTS: none

CATALOG NUMBER: 33 ARTIFACT NUMBER: PR85 1026-11

EXCAVATION: Hamilton

WEIGHT TYPE: dome weight

MARKS: 2 marks on the shoulder - not distinguishable

ORIGIN: English

ILLUSTRATIONS:

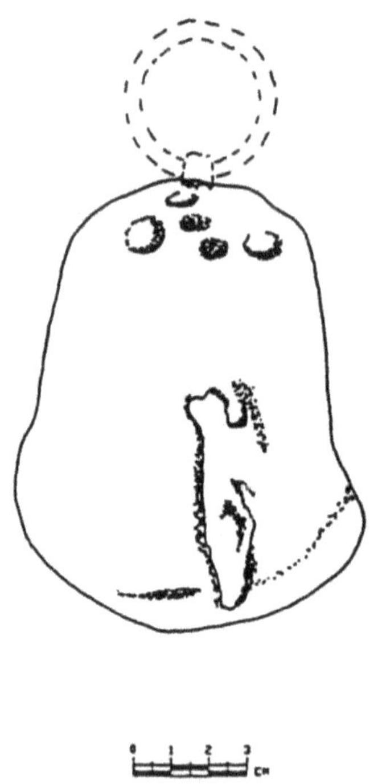

DESCRIPTION: The shape of this weight is a standard bell form. On the side of the body of the weight, a section of lead has been altered, either from adjusting the weight or from repairs made from faulty casting. The weight had an iron ring on top as evidenced by two holes containing remnants of corroded iron. On the shoulders of the weight are two marks which are not distinguishable.

DIMENSIONS: HEIGHT (cm): 8.90
DIAMETER/BASE (cm): 7.62

WEIGHT IN KILOS: 3.18 WEIGHT IN POUNDS: 7.01

PROBABLE WEIGHT: 9 lb A. (with handle and staple) VARIANCE: -22.3%

PARALLELS: This bell weight is similar to PR85 1034-9.

MATERIALS: lead body - iron ring

COMMENTS: used on balance scales and beams - could be used on a scale pan or on a hook - for larger objects

CATALOG NUMBER: 34 ARTIFACT NUMBER: PR85 1026-11.1

EXCAVATION: Hamilton

WEIGHT TYPE: dome weight

MARKS: none

ORIGIN: English

ILLUSTRATIONS:

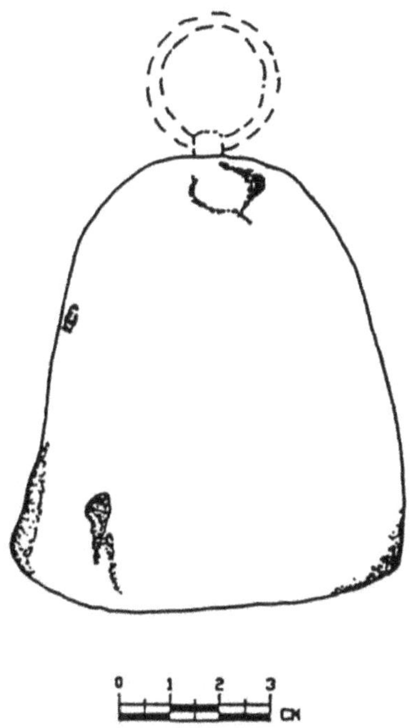

DESCRIPTION: This weight is crudely made and there is the indentation of a possible assayers mark on the shoulder (not distinguishable). The weight had an iron ring or handle on top.

DIMENSIONS: HEIGHT (cm): 8.90
DIAMETER/BASE (cm): 7.90

WEIGHT IN KILOS: 3.18 WEIGHT IN POUNDS: 7.0

PROBABLE WEIGHT: 9 lb A. (with handle and staple) VARIANCE: -22.3%

PARALLELS: PR85 1026-13 is similar to PR89 859-12.

MATERIALS: lead body - iron ring

COMMENTS: The top of the weight has been extensively worn so that identification of the mark is not possible.

CATALOG NUMBER: 35 ARTIFACT NUMBER: PR85 1034-9

EXCAVATION: Hamilton

WEIGHT TYPE: dome weight

MARKS: none

ORIGIN: unknown

ILLUSTRATIONS:

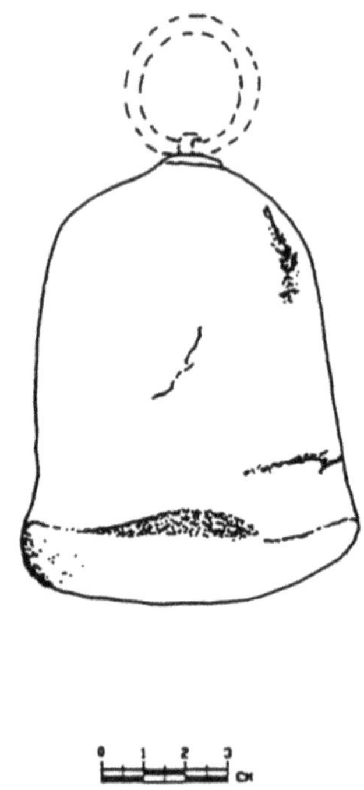

DESCRIPTION: This weight does not have any seals or marks on its shoulders and is possibly not a sanctioned weight even though it appears to conform to the avoirdupois standard. The weight would have had an iron ring handle and if the weight did conform to the standard, the ring would have had to have weighed .23 kilos or one half pound.

DIMENSIONS: HEIGHT (cm): 9.23
DIAMETER/BASE (cm): 7.30

WEIGHT IN KILOS: 2.95 WEIGHT IN POUNDS: 6.50

PROBABLE WEIGHT: 7 lb A. (with handle and staple) VARIANCE: -7.2%

PARALLELS: PR89 859-12 is similar PR85 1026-11.

MATERIALS: lead body - iron ring

COMMENTS: with the ring, this weight was probably seven pounds by the Imperial system - since there were assayers marks or regis marks, it is possible that the weight was either foreign or from a small shop some distance from London.

CATALOG NUMBER: 36 ARTIFACT NUMBER: PR89 859-12

EXCAVATION: Hamilton

WEIGHT TYPE: dome weight

MARKS: - 2 "WB" marks on the shoulder of the weight - possibly an assayers mark

ORIGIN: English

ILLUSTRATIONS:

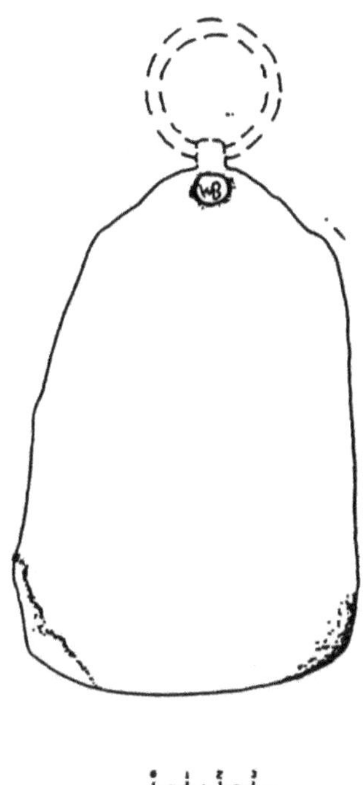

DESCRIPTION: This weight has the remnants of an iron ring mounted on it's top as well as two legible "WB" marks.

DIMENSIONS: HEIGHT (cm): 16.20
DIAMETER/BASE (cm): 10.50

WEIGHT IN KILOS: 13.64 WEIGHT IN POUNDS: 30.07

PROBABLE WEIGHT: 35 lb A. (with handle and staple) VARIANCE: -14.10%

PARALLELS: PR89 859-12 has a similar body shape to PR85 1034-9.

MATERIALS: lead body - iron ring

COMMENTS: The iron ring for this weight probably had a weight of five pounds, making the entire weight thirty-five pounds by the avoirdupois standard.

CATALOG NUMBER: 37 ARTIFACT NUMBER: PR90 843-5

EXCAVATION: Hamilton

WEIGHT TYPE: brick weight

MARKS: none

ORIGIN: unknown

ILLUSTRATIONS:

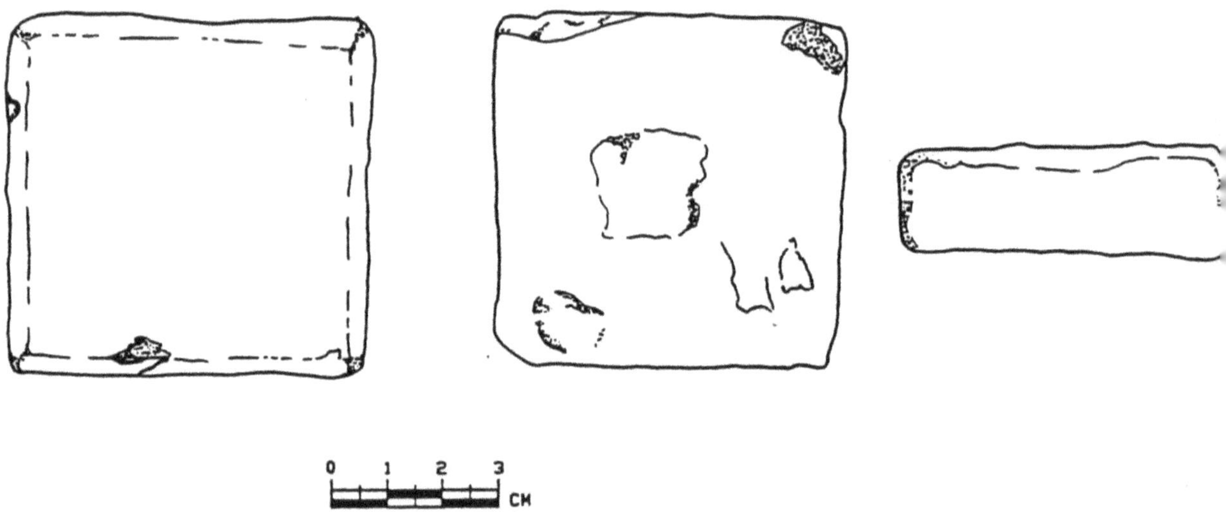


DESCRIPTION: This weight is square in shape and the upper edges of the weight have been slightly beveled. The weight was formed by pouring lead into a mold, resulting in crazing on the bottom surface, which was exposed to the air while cooling. The weight contains no marks.

DIMENSIONS: HEIGHT (cm): 2.00
WIDTH/BASE (cm): 6.40
LENGTH/BASE (cm): 6.20

WEIGHT IN KILOS: 1.30 WEIGHT IN POUNDS: 2.87

PROBABLE WEIGHT: 3 lb A. VARIANCE: -4.47%

PARALLELS: none

MATERIALS: lead

COMMENTS: This is a simple, multi purpose form of weight.

CATALOG NUMBER: 38 ARTIFACT NUMBER: PR89 857-9

EXCAVATION: Hamilton

WEIGHT TYPE: brick weight

MARKS:
- "LV" - fifty-five pounds
- "SP" - owners mark
- partially legible "C" mark

ORIGIN: unknown

ILLUSTRATIONS:

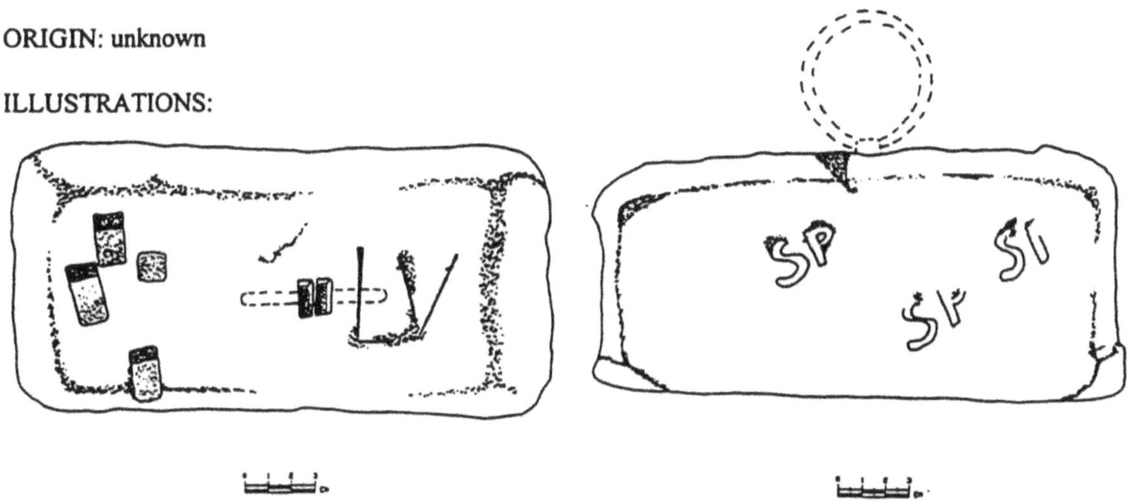


DESCRIPTION: Two centrally located holes on the top surface of this weight indicates that the weight probably had an iron handle. Three additional holes located at one end of the top surface contain iron oxide residue indicating that iron nails or plugs had been added as a means of altering the weight. The ring handle would have been fastened using a large staple.

DIMENSIONS: HEIGHT (cm): 11.40
LENGTH (cm): 21.50
WIDTH (cm): 10.50

WEIGHT IN KILOS: 24.4 WEIGHT IN POUNDS: 53.79

PROBABLE WEIGHT: 55-57 lbs A. (with handle and staple) VARIANCE: -2.2%

PARALLELS: none

MATERIALS:lead - iron ring handle

COMMENTS: The "LV" inscription, appears to be an accurate indicator of the intended mass of this weight. Weights such as this were common for measuring bushels of wheat and were in popular use until 1800 AD.

CATALOG NUMBER: 39 ARTIFACT NUMBER: PR89 859-11

EXCAVATION: Hamilton

WEIGHT TYPE: brick weight

MARKS: - "VXX" inscribed on the side
- 2 - fleur-de-lys over boats

MONARCH: possibly Charles I ORIGIN: English

ILLUSTRATIONS:

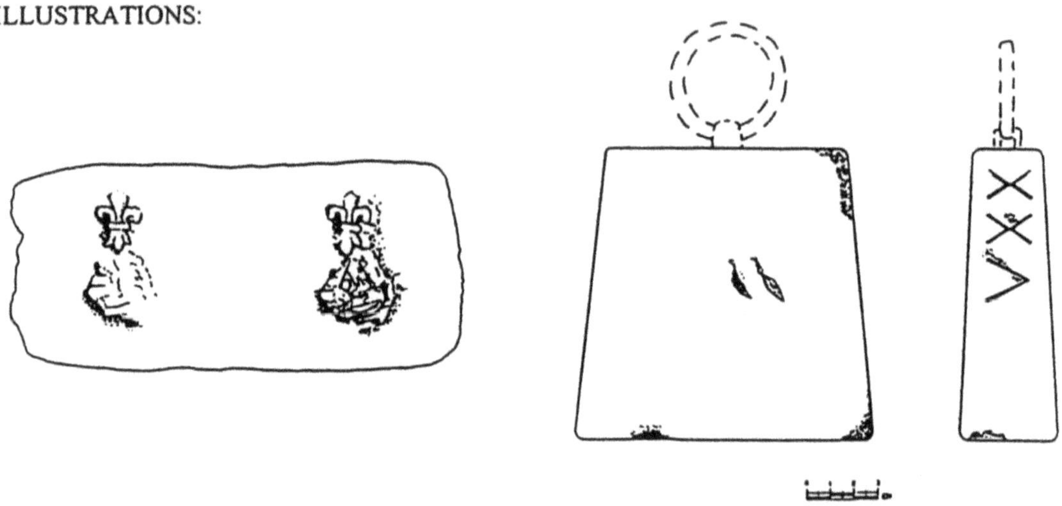

DESCRIPTION: This weight originally had an iron ring handle mounted on the top surface with the mark of a fleur-de-lys over a ship on each side of the handle. The earliest known mark which is similar to those found on the weight appears on an English coin weight dated circa 1650 (Houben 1978, p.7)

DIMENSIONS: HEIGHT (cm): 16.80
WIDTH/BASE (cm): 11.40
THICK/BASE (cm): 4.70

WEIGHT IN KILOS: 11.25 WEIGHT IN POUNDS: 24.8

PROBABLE WEIGHT: 25 lb A. (with ring and handle) VARIANCE: -.08%

PARALLELS: similar to PR 859-13.

MATERIALS: lead body, iron ring handle

COMMENTS: used to measure wheat on smaller pan scales

CATALOG NUMBER: 40 ARTIFACT NUMBER: PR89 859-13

EXCAVATION: Hamilton

WEIGHT TYPE: brick weight

MARKS: - 2 fleur-de-lys over a ship with sails, located one on each side of an iron ring handle

MONARCH: possibly Charles I ORIGIN: England

ILLUSTRATIONS:

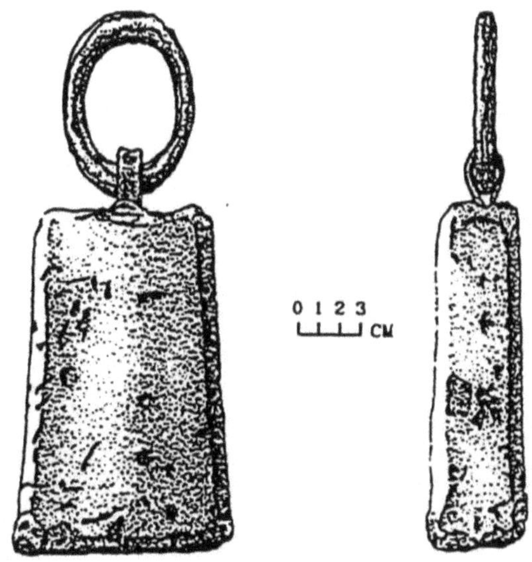


DESCRIPTION: Similar in shape to PR89 859-11, this weight has a reconstructed ring handle taken from the mold of an encrustation. Unlike 859-11 however, this weight does not have any weight designations inscribed on it.

DIMENSIONS: HEIGHT (cm): 12.00
WIDTH/BASE (cm): 9.50
THICK/BASE (cm): 4.30

WEIGHT IN KILOS: 6.13 WEIGHT IN POUNDS: 13.54

PROBABLE WEIGHT: 14 lb A. (with handle and staple) VARIANCE: -3.3%

PARALLELS: This weight is similar to PR89 859-11.

MATERIALS: lead body - iron ring handle

COMMENTS: As a multiple of the 7 lb standard used for measuring wheat, this weight would have been useful in a bakery or for measuring grain for sale.

CATALOG NUMBER: 41 ARTIFACT NUMBER: N.P.

EXCAVATION: Hamilton

WEIGHT TYPE: pai weight

MARKS: - Archangel Michael - Plumbers Guild stamp or mark of the shop of Messrs. De Grave and Company or Thomas Overing
- dagger - City of London
- Crown over "C" - Charles I or II

MONARCH: 17nth century ORIGIN: English

ILLUSTRATIONS:

Enlarged drawing of the Michael figure.

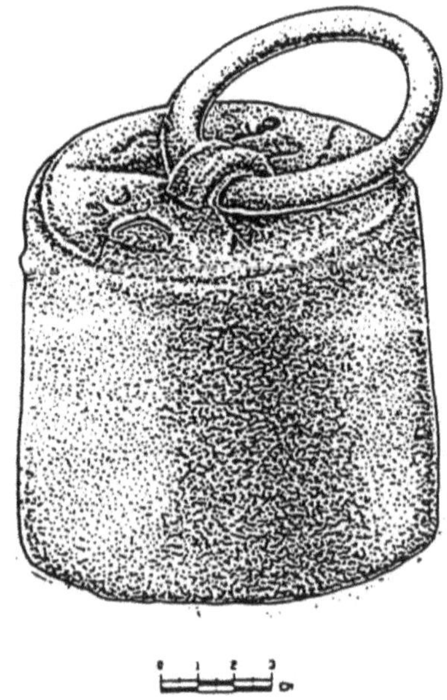

DESCRIPTION: The top markings on this weight are very interesting and include some of the best preserved and diagnostic marks recovered. The body of the weight was formed by pouring molten lead into a cast iron, pail-shaped container. An iron ring handle was then placed on the top of the weight using a staple, which fastened it to the weight as the metal cooled.

DIMENSIONS: HEIGHT (cm): 9.82
DIAMETER/TOP(cm): 8.90
DIAMETER.BASE(cm): 11.43

WEIGHT IN KILOS: 6.1 WEIGHT IN POUNDS: 13.51

PROBABLE WEIGHT: 14 lb A. (with handle and staple) VARIANCE: -3.5%
MATERIALS: iron exterior/lead interior/iron handle

COMMENTS: possibly used for weighing wheat

CATALOG NUMBER: 42 ARTIFACT NUMBER: PR85 1026-13

EXCAVATION: Hamilton

WEIGHT TYPE: pail weight

MARKS:
- Archangel Michael - the Plumbers Guild stamp
- dagger - City of London
- Crown over "C" - Charles I or II

MONARCH: 17nth century English ORIGIN: English

ILLUSTRATION:

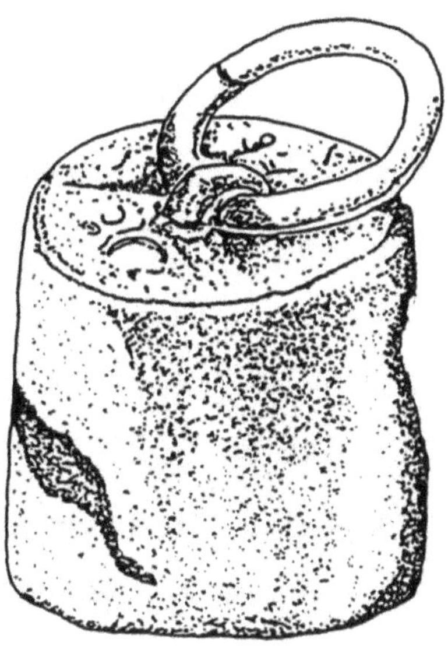

DESCRIPTION: Similar to PR N.P.,this weight has well preserved stamps on its top surface. The exterior iron surface of the body of this weight has exfoliated to a great degree and the iron ring handle and staple have been reproduced by means of a Hysol epoxy cast.

DIMENSIONS: HEIGHT (cm): 11.20
DIAMETER/TOP (cm): 10.10
DIAMETER/BASE (cm): 12.89

WEIGHT IN KILOS: 6.136 WEIGHT IN POUNDS: 13.53

PROBABLE WEIGHT: 14 lb A. (with handle and staple) VARIANCE: -3.4%

MATERIALS: iron casing, lead core

COMMENTS: Marks on this weight have survived extremely well because the iron ring handle protected the top surface of the weight and the marks were stamped into a lead surface which has resisted corrosion. This weight could have been used to weigh wheat.

CATALOG NUMBER: 43 ARTIFACT NUMBER: PR82 129.6

EXCAVATION: Hamilton

WEIGHT TYPE: composite weight

MARKS: none

ORIGIN: English

ILLUSTRATIONS:

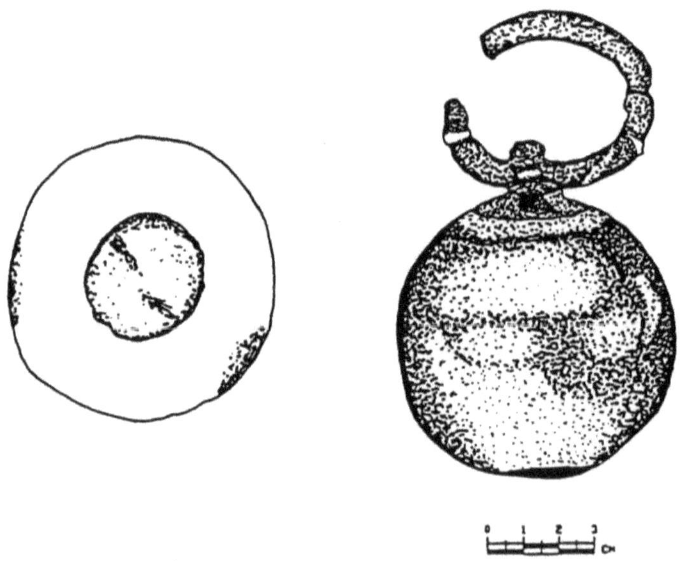

DESCRIPTION: The weight had an iron ring handle which was reconstructed from the concretion using epoxy. The body of the weight is circular and was made by pouring lead into a round brass container. The iron ring handle was attached to the body of the weight by means of an iron staple.

DIMENSIONS: HEIGHT (cm): 12.00
DIAMETER (cm): 7.50

WEIGHT IN KILOS: 2.14 WEIGHT IN POUNDS: 4.72

PROBABLE WEIGHT: 5 lb A. VARIANCE: -5.6%

PARALLELS: Coin and counterpoise weight were very similar in shape (Kisch 1965, p.62).

MATERIALS: iron ring handle - lead body with brass covering

COMMENTS: This type of weight was clearly to be suspended during use and would likely have been used on steelyards for weighing larger objects, and also on coin weights.
Kisch illustrates steelyards from Pompeii which, although they were shaped in the form of human heads, would have performed the same functions as the round models (Kisch, 1965, p. 59). A coin weight similar to PR82 129-6 is illustrated on page sixty-three (Kisch 1965).

CATALOG NUMBER: 44 ARTIFACT NUMBER: PR83 423-40

EXCAVATION: Hamilton

WEIGHT TYPE: composite weight

MARKS: none

ORIGIN: unknown

ILLUSTRATIONS:

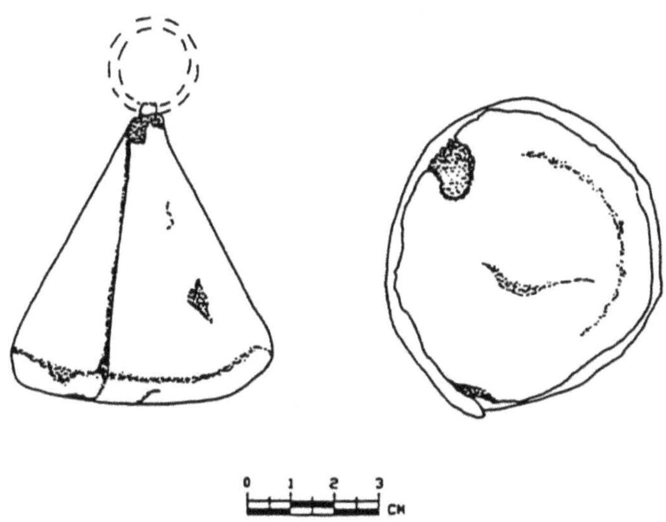


DESCRIPTION: PR83 423-40 is a conical shaped weight constructed with a lead core covered by a copper sheet with one seam extending from the top to the bottom of the weight. The base of the weight is exposed lead. Traces of iron oxide indicate that this weight had an iron handle extending from the top of the cone.

DIMENSIONS: HEIGHT (cm): 5.72
DIAMETER/BASE (cm): 5.72

WEIGHT IN KILOS: .32 WEIGHT IN POUNDS: .71

PROBABLE WEIGHT: 1 lb A.(with handle and staple) VARIANCE: -29%

PARALLELS: This weight is very similar to PR305.

MATERIALS: lead body - copper layer

COMMENTS: Conical weights were commonly used on balance beams and scales since they could either be hung from a hook or placed on a pan.

CATALOG NUMBER: 45 ARTIFACT NUMBER: PR 87 426-6

EXCAVATION: Hamilton

WEIGHT TYPE: composite weight

MARKS: none

ILLUSTRATION:

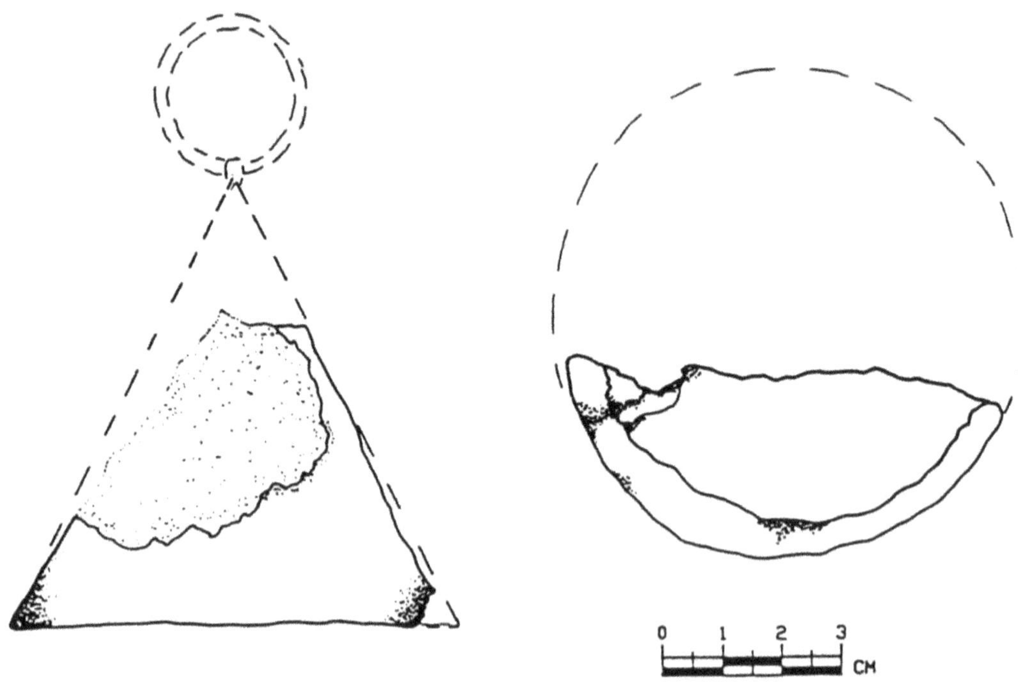


DESCRIPTION: This weight appears to have been a more common cone weight form. The side surfaces of this weight were badly deteriorated but dimensions and details were taken from the field notes.

DIAMETER/BASE (cm): 7.0

WEIGHT IN KILOS: not determined

PARALLELS: This weight is similar to PR83 423-40, and PR90 897-9.1 in shape and size.

MATERIALS: copper - bound aggregate material added for mass

COMMENTS: The tip of this weight was not recovered but it is most probable that the weight had an iron ring handle and staple. With its tip in place, this weight would have been almost exactly as wide as it was tall, which is a common dimensional feature of this weight type.

CATALOG NUMBER: 46 ARTIFACT NUMBER: PR87 414-8

EXCAVATION: Hamilton

WEIGHT TYPE: composite weight

MARKS: none

ILLUSTRATION:

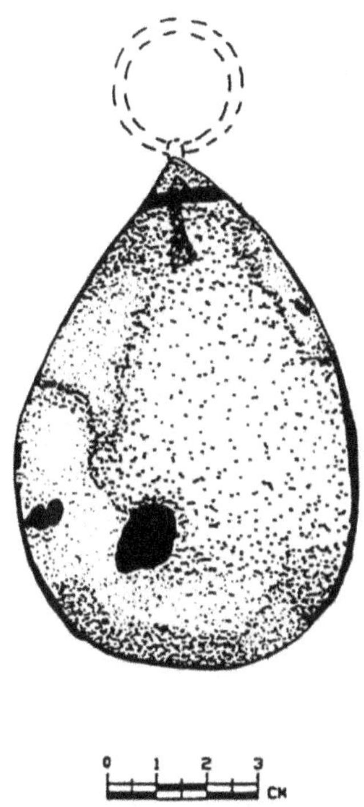


DESCRIPTION: This weight was largely constructed from x-rays since the concretion did not survive the trip to College Station. The weight had a unique tear-drop form and consisted of a thin shell (possibly copper) which appears to have been filled with sand or bound aggregate material.

DIMENSIONS: HEIGHT (cm): 10.0
DIAMETER/WIDEST POINT (cm): 7.0

WEIGHT IN KILOS: not determined

PARALLELS: This is a unique weight. X-rays indicate that the weight had an iron ring handle, similar to PR90 897-9.1.

MATERIALS: Staining on small pieces of the remaining concretion indicate that the weight case was possibly made of copper -

COMMENTS: A tear drop shaped weight such as this would have been suspended from a beam or scale for measurement.

CATALOG NUMBER: 47 ARTIFACT NUMBER: PR89 305

EXCAVATION: Hamilton

WEIGHT TYPE: composite weight

MARKS: none

ORIGIN: unknown

ILLUSTRATIONS:

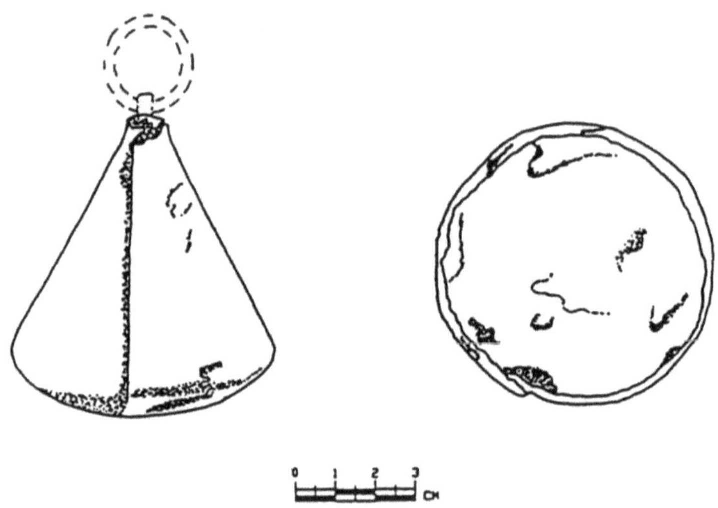

DESCRIPTION: This weight is conical in form with a single seamed copper sheet cover, over a lead core. The base of the weight is not enclosed and the exposed lead is pitted and rough in texture. The weight originally had an iron handle, used for suspending it from a scale or beam.

DIMENSIONS: HEIGHT (cm): 6.35
DIAMETER/BASE (cm): 6.35

WEIGHT IN KILOS: .575 (less handle and staple) WEIGHT IN POUNDS: 1.27

PROBABLE WEIGHT: 1 lb 6 oz (with handle) VARIANCE: -15.34

PARALLELS: PR 305 and PR83 423-40 are almost identical in style and construction.

MATERIALS: lead covered by copper sheeting - a single seam down the side

COMMENTS: Conical weights were commonly used on balance scales since they could either be hung from a hook or placed on a pan.

CATALOG NUMBER: 48 ARTIFACT NUMBER: PR90 897-9.1

EXCAVATION: Hamilton

WEIGHT TYPE: composite weight

MARKS: none

MONARCH: 17 century English

ILLUSTRATION:

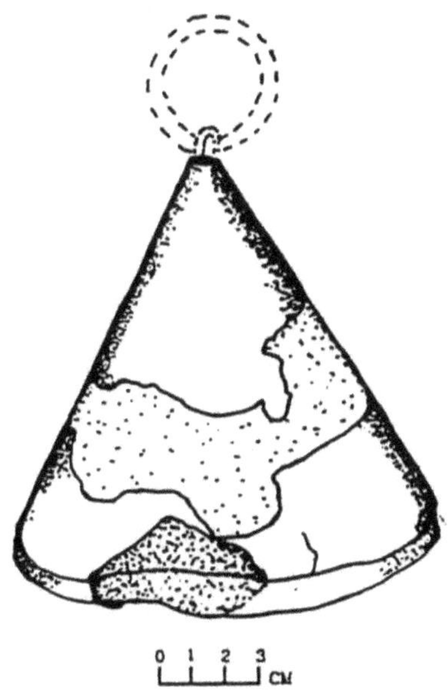


DESCRIPTION: This weight was part of a large concretion that was recovered from the hearth area in building five. It is very unique since it was a very thin shell made of copper and was filled with bound aggregate material to add mass. A thick plate, possibly wood was crimped to the sides of the cone to contain sand-like core of the weight.

DIMENSIONS: HEIGHT (cm): 10.140
DIAMETER/BASE (cm): 9.94

WEIGHT IN KILOS: not determined

PARALLELS: Very similar in form to PR89 305

MATERIALS: The weight was constructed by filling an iron casing with bound aggregate material and crimping a bottom plate into place to form a bottom.

COMMENTS: Sand, possibly mixed with a binding agent was used to add mass to the weight instead of lead.

CATALOG NUMBER: 49

ARTIFACT NUMBER: N.P.

EXCAVATION: Marx

WEIGHT TYPE: disc weight

MARKS:
- "LS" Owners mark
- dagger (city of London)
- Ā - Avoirdupois standard
- crown over orb - legal weight
- Founders Company Mark

ORIGIN: English

ILLUSTRATIONS:

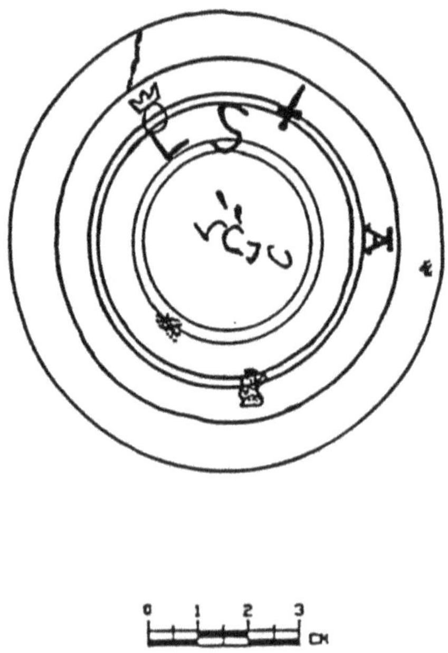

DESCRIPTION: Centrally located on the top surface of the weight is a mark which is not distinguishable. The mark of the Founders Company guild is a ewer and on this weight the handle is on the left. The "A" signifies the reign of Queen Anne.

DIMENSIONS: HEIGHT (cm): .90
DIAMETER/BASE (cm): 8.0

WEIGHT IN KILOS: .196

WEIGHT IN POUNDS: .432

PROBABLE WEIGHT: 8 oz A.

VARIANCE: -13.6%

PARALLELS: very common style and markings.

MATERIALS: bronze

COMMENTS: Disc weights of this style were used on balance beams for all types of measurement.

CATALOG NUMBER: 50 ARTIFACT NUMBER: PR 355

EXCAVATION: Marx

WEIGHT TYPE: disc weight

MARKS:
- crown over C - 17nth Century (Charles I&II)
- dagger - city of London
- Founders Company Mark

MONARCH: Charles I or II ORIGIN: England

ILLUSTRATIONS:

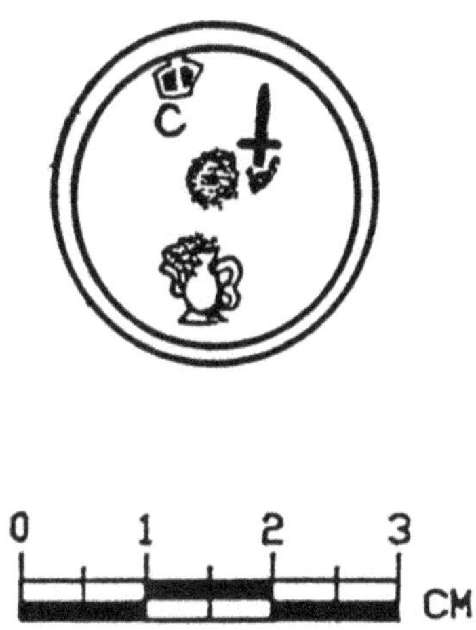


DESCRIPTION: This is a very small disc weight which was probably used for measuring precious metals and coins.

DIMENSIONS: HEIGHT (cm): .38
DIAMETER/BASE (cm): 2.54

WEIGHT IN KILOS: .12 WEIGHT IN POUNDS: .265

PROBABLE WEIGHT: 4 oz A. VARIANCE: +6%

PARALLELS: Similar in size to SP1. S5.2, found at the St. Peters Church excavation.

MATERIALS: bronze

COMMENTS: none

CATALOG NUMBER: 51

ARTIFACT NUMBER: PR 156

EXCAVATION: Marx

WEIGHT TYPE: disc weight

MARKS:
- dagger - City of London
- Founders Company guild mark
- shield with cross - undetermined
- A - avoirdupois standard

ORIGIN: English

ILLUSTRATIONS:

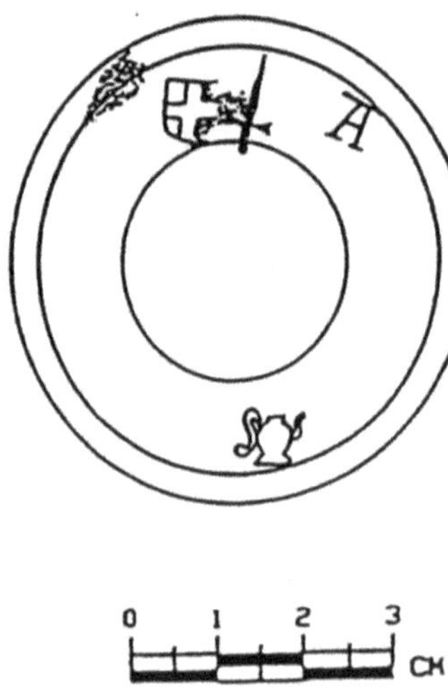


DESCRIPTION: This is a medium sized disc weight which could have been used on a balance scale.

DIMENSIONS: HEIGHT (cm): .75
DIAMETER/BASE (cm): 5.20

WEIGHT IN KILOS: .11

WEIGHT IN POUNDS: .243

PROBABLE WEIGHT: 4 oz A.

VARIANCE: -2.8%

PARALLELS: This weight is similar in size to PR 743.

MATERIALS: bronze

COMMENTS: The dagger was the earliest insignia for the City of London and the use of the crest followed later. After 1826, the dagger was always included in the coat of arms.

CATALOG NUMBER: 52

ARTIFACT NUMBER: D93-1

EXCAVATION: Marx

WEIGHT TYPE: disc weight

MARKS:
- "LS" - owners marks
- dagger - City of London
- crown over orb - legal weight
- orb - added after assaying
- A - avoirdupois standard

ORIGIN: England

ILLUSTRATIONS:

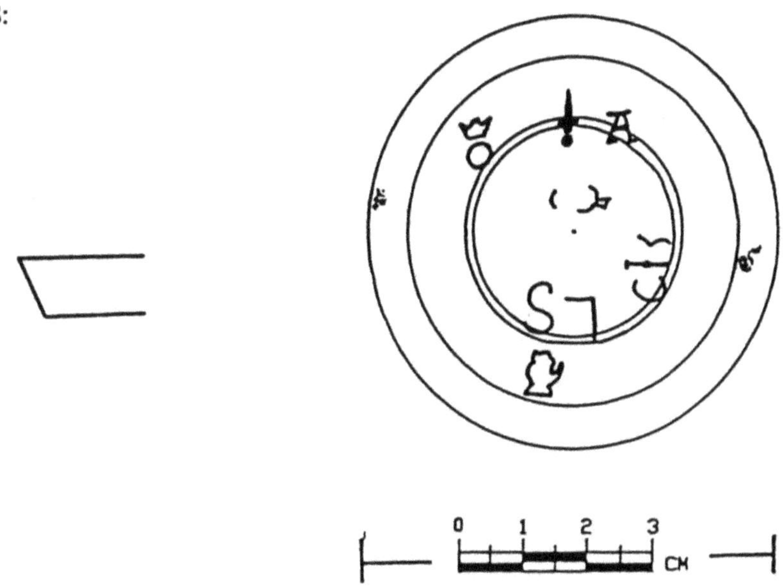


DESCRIPTION: This is a medium sized disc weight, compared to other weights from the site and it was manufactured under the controls of the Founders Company in the City of London.

DIMENSIONS: HEIGHT (cm): .90
DIAMETER/BASE (cm): 6.45

WEIGHT IN KILOS: .198

WEIGHT IN POUNDS: .437

PROBABLE WEIGHT: 8 oz A.

VARIANCE: -12.6%

PARALLELS: This weight is similar in design Marx (N.P.).

MATERIALS: bronze

COMMENTS: none

CATALOG NUMBER: 53 ARTIFACT NUMBER: D93-5

EXCAVATION: Marx

WEIGHT TYPE: disc weight

MARKS:
- "C" in a box - 17nth cent. (Charles I or II)
- dagger - City of London
- Ā - avoirdupois standard

MONARCH: 17nth century ORIGIN: England

ILLUSTRATIONS:

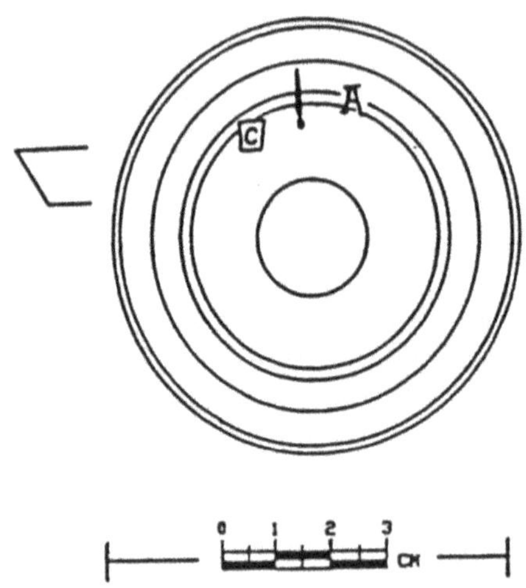


DESCRIPTION: This is a medium sized weight which shows no signs of having carried the mark of the Founders Company of London. Because the "C" mark differs from the Founders Company guild weights, it is possible that this was not a legal weight by the king's standard or that the weight was officially sanctioned and simply made at a different location.

DIMENSIONS: HEIGHT (cm): 1.55
DIAMETER/BASE(cm): 7.75

WEIGHT IN KILOS: .22 WEIGHT IN POUNDS: .49

PROBABLE WEIGHT: 8 oz A. VARIANCE: -2%

PARALLELS: The monarch's mark or regis mark is different than any other mark from the site.

MATERIALS: bronze

COMMENTS: none

CATALOG NUMBER: 54 | ARTIFACT NUMBER: D93-6

EXCAVATION: Marx

WEIGHT TYPE: disc weight

MARKS:
- crown over C - Charles I or II
- dagger - City of London
- "LS" - possibly an owners mark
- A - avoirdupois standard

MONARCH: 17nth century | ORIGIN: England

ILLUSTRATIONS:

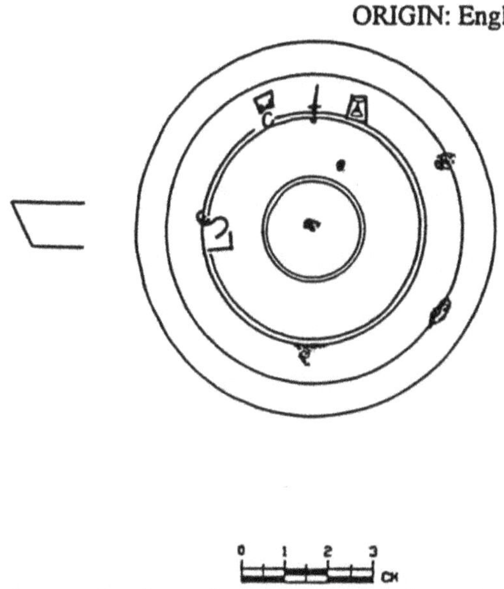

DESCRIPTION: This weight is a medium-large disc weight which shows signs of heavy use-wear on it's sides.

DIMENSIONS: HEIGHT (cm): 1.27
DIAMETER/TOP (cm): 8.10
DIAMETER/BASE(cm): 7.75

WEIGHT IN KILOS: .42 | WEIGHT IN POUNDS: .93

PROBABLE WEIGHT: 1 lb A. | VARIANCE: -7%

PARALLELS: The weight is similar to D93-1.

MATERIALS: bronze

COMMENTS: The "LS" mark has been very crudely inscribed on the top surface, compared to the other marks suggesting that it was an owners mark.

CATALOG NUMBER: 55

ARTIFACT NUMBER: PR 93-7

EXCAVATION: Marx

WEIGHT TYPE: disc weight

MARKS:
- "CR" - probably Charles Regis (Chas. I or II)
- dagger - City of London
- Founders Company guild Ewer
- cross and orb - legal weight (assayers mark)
- "LS" - possibly owners mark

MONARCH: 17nth century

ORIGIN: England

ILLUSTRATIONS:

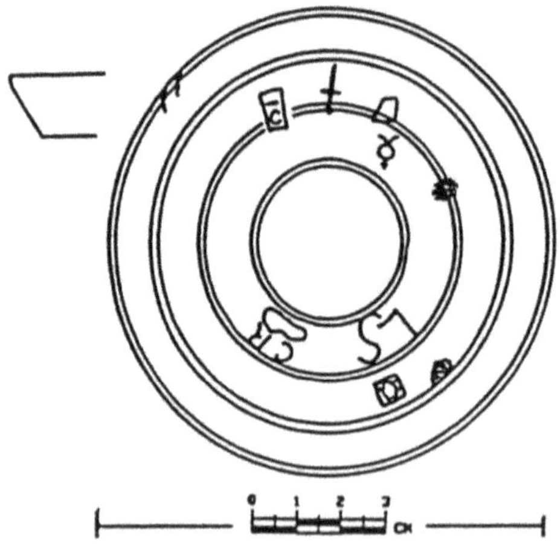

DESCRIPTION: PR 93-7 is one of the larger disc weights recovered from the site and shows extensive signs of abrasion and wear on it's sides.

DIMENSIONS: HEIGHT (cm): 1.55
DIAMETER/TOP (cm): 7.75
DIAMETER/BASE(cm): 7.45

WEIGHT IN KILOS: .49

WEIGHT IN POUNDS: 1.08

PROBABLE WEIGHT: 1 lb A.

VARIANCE: +8%

PARALLELS: The weight is similar in body shape and size to D95-1 with the substitution of the crown over C for the cross over orb.

MATERIALS: bronze

COMMENTS: Similar to D93-6, the "LS" mark is very crude suggesting that it was an owners mark.

CATALOG NUMBER: 56 ARTIFACT NUMBER: D95-1

EXCAVATION: Marx

WEIGHT TYPE: disc weight

MARKS:
- crown over "C" - Charles I or II
- dagger - City of London
- Founders Company guild ewer
- "GS" inscribed on upper rim - owners mark
- A - avoirdupois standard

MONARCH: 17nth century ORIGIN: England

ILLUSTRATIONS:

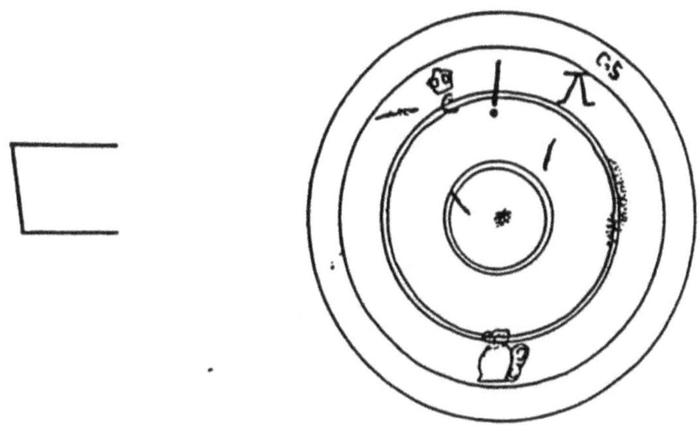

DESCRIPTION: This is a medium sized disc weight and the inscribed "regis mark" (crown over "C") is a very faint mark.

DIMENSIONS: HEIGHT (cm): 1.40
DIAMETER/TOP(cm): 7.62
DIAMETER. BASS(cm): 6.98

WEIGHT IN KILOS: .43 WEIGHT IN POUNDS: .95

PROBABLE WEIGHT: 1 lb A. VARIANCE: -5%

PARALLELS: This weight is similar to D93-5.

MATERIALS: bronze

COMMENTS: none

CATALOG NUMBER: 57 ARTIFACT NUMBER: D95-2

EXCAVATION: Marx

WEIGHT TYPE: disc weight

MARKS:
- crown over "C" - Charles I or II
- dagger - City of London
- Founders Company guild mark
- "GS" inscribed on rim - owners mark
- A - avoirdupois standard

MONARCH: 17nth century ORIGIN: England

ILLUSTRATIONS:

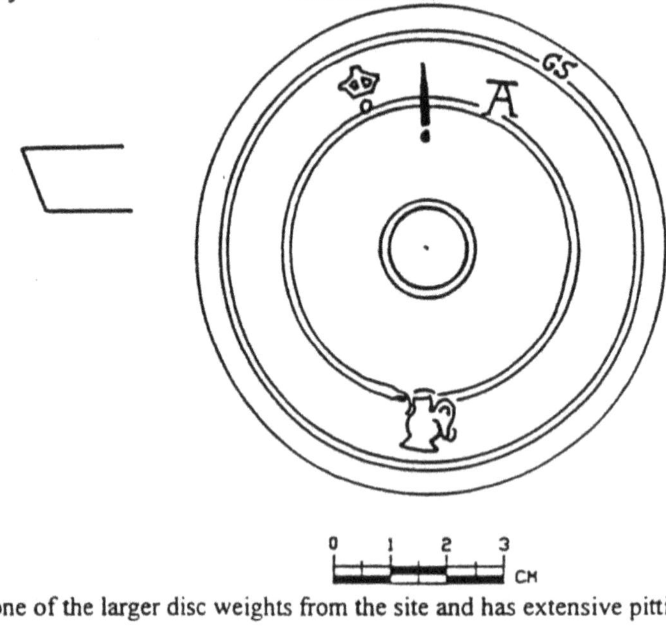


DESCRIPTION: This is one of the larger disc weights from the site and has extensive pitting on the upper surface.

DIMENSIONS: HEIGHT (cm): 1.27
DIAMETER/TOP (cm): 7.62
DIAMETER/BASE(cm): 6.98

WEIGHT IN KILOS: .42 WEIGHT IN POUNDS:.93

PROBABLE WEIGHT: 1 lb A. VARIANCE: -7%

PARALLELS: D95-2 is very similar to D95-1.

MATERIALS: bronze

COMMENTS: none

CATALOG NUMBER: 58

ARTIFACT NUMBER: D95-3

EXCAVATION: Marx

WEIGHT TYPE: disc weight

MARKS:
- crown over "C" - Charles I or II
- dagger - City of London
- Founders company guild ewer
- A - avoirdupois standard

MONARCH: 17nth century

ORIGIN: England

ILLUSTRATIONS:

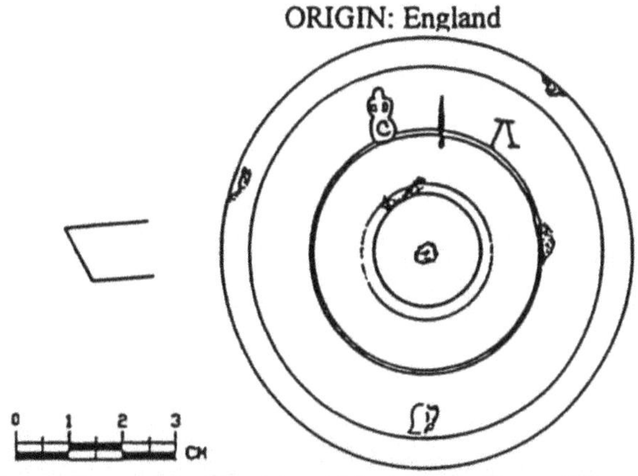


DESCRIPTION: D95-3 is a large disc weight. The top and bottom surfaces of the weight are heavily abraded and the Founders Company guild ewer and the regis mark are very faint.

DIMENSIONS: HEIGHT (cm): 1.40
DIAMETER/TOP(cm): 7.62
DIAMETER/BASE(cm): 7.30

WEIGHT IN KILOS: .435

WEIGHT IN POUNDS: .96

PROBABLE WEIGHT: 1 lb A.

VARIANCE: -4%

PARALLELS: While D93-5 has no owners mark, it is strikingly similar to D95-2.

MATERIALS: bronze

COMMENTS: Both D95-3 and D95-2 are probably part of the same set of pan scale weights.

CATALOG NUMBER: 59

ARTIFACT NUMBER: PR95-4

EXCAVATION: Marx

WEIGHT TYPE: disc weight

MARKS:
- incised "regis" mark - similar to D95-3 but indistinguishable
- dagger - City of London
- Ewer - possibly the Founders Company guild mark
- "L" - probable owners mark
- A - avoirdupois standard

MONARCH: Possibly 17nth century

ORIGIN: English

ILLUSTRATIONS:

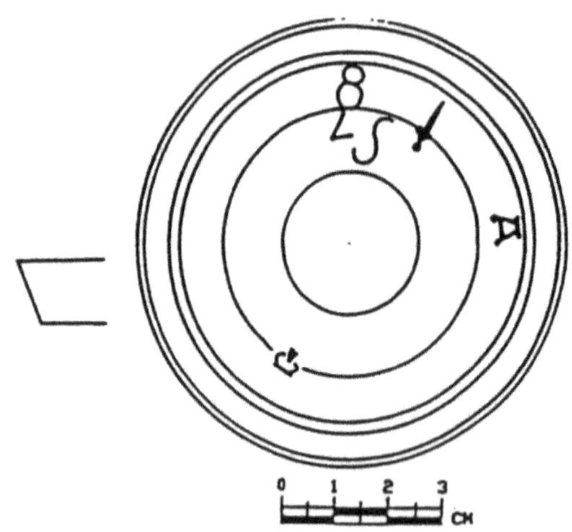


DESCRIPTION: This is a large weight which in addition to having the same markings as other weights, has an impression in the shape of a figure eight.

DIMENSIONS: HEIGHT (cm): 1.27
DIAMETER/TOP(cm): 7.85
DIAMETER/BASE(cm): 7.62

WEIGHT IN KILOS: .42

WEIGHT IN POUNDS: .93

PROBABLE WEIGHT: 1 lb A.

VARIANCE: -7%

PARALLELS: This weight is most similar to D93-1.

MATERIALS: bronze

COMMENTS: The "figure eight" mark is probably the remains of a "regis" mark similar to that of Charles II, on D95-3.

CATALOG NUMBER: 60 ARTIFACT NUMBER: NP

EXCAVATION: Marx

WEIGHT TYPE: disc weight

MARKS:
- Crown over "C"- possible Charles I
- dagger - City of London
- Founders Company guild ewer - right facing handle
- A - avoirdupois standard

MONARCH: early 17nth century ORIGIN: English

ILLUSTRATIONS

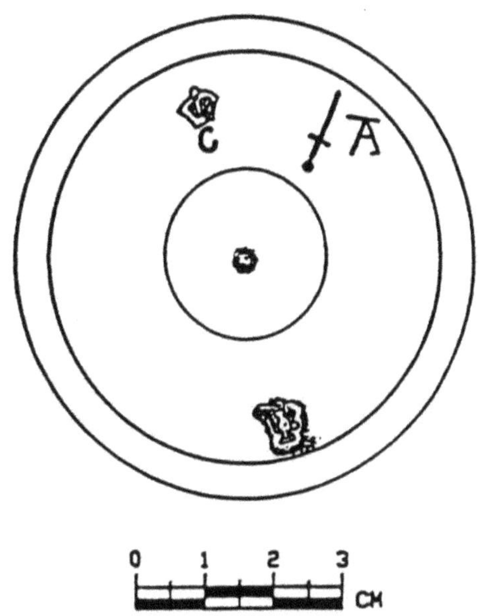


DESCRIPTION: All marks on this weight are legible and the conserved weight has a very glossy, bronze finish. The taper of the weight from top to bottom is identical to that of weight number D95-1.

DIMENSIONS: HEIGHT (cm): 1.10
DIAMETER/TOP(cm): 6.35
DIAMETER/BASE(cm): 6.00

WEIGHT IN KILOS: .199 WEIGHT IN POUNDS: .44

PROBABLE WEIGHT: 8 oz A. VARIANCE: -12%

PARALLELS: This weight is very similar to PR85 1024-7.5) and D95-1.

MATERIALS: bronze

COMMENTS: none

CATALOG NUMBER: 61 ARTIFACT NUMBER: SP.1.S51.

EXCAVATION: Mayes

WEIGHT TYPE: coin or disc weight

MARKS: large "M" on each side

ORIGIN: English

ILLUSTRATIONS:

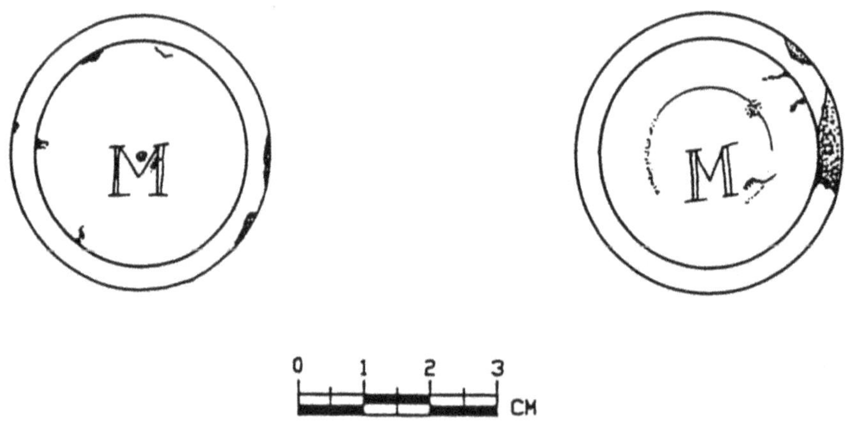


DESCRIPTION: This weight is similar in design to the other disc weights except that it has a large "M" inscribed on it's top and bottom surfaces.

DIMENSIONS: HEIGHT (cm): .64
DIAMETER/BASE (cm): 4.15

WEIGHT IN KILOS: .05 WEIGHT IN POUNDS: .11

PROBABLE WEIGHT: 2 oz A. VARIANCE: -12%

PARALLELS: none

MATERIALS: bronze

COMMENTS: While the "M" mark is possibly the mark of Queen Mary II (1688-1694), it is more probable that the mark is the mark of the first Mexican mint, which made weights between 1553 and 1559.

CATALOG NUMBER: 62

ARTIFACT NUMBER: SP.1 S5.2

EXCAVATION: St. Peters Church, Mayes

WEIGHT TYPE: disc weight

MARKS:
- crown over "W" = King William III
- dagger - City of London
- Founders Company guild ewer
- A - avoirdupois standard

MONARCH: King William

ORIGIN: England

ILLUSTRATIONS:

DESCRIPTION: SP.1 S5.2 is a small disc weight commonly used for money weights and on small balance beams. The top surface of the weight is heavily pitted.

DIMENSIONS: HEIGHT (cm): .45
DIAMETER/BASE (cm): 3.17

WEIGHT IN KILOS: .245

WEIGHT IN POUNDS: .53

PROBABLE WEIGHT: 8 oz A.

VARIANCE: +6%

PARALLELS: The weight is most similar to PR 355.

MATERIALS: bronze

COMMENTS: The crown over "W" mark is partially obscured.

CATALOG NUMBER: 63 ARTIFACT NUMBER: NS.2. A7A.1d

EXCAVATION: New Street, Priddy

WEIGHT TYPE: disc weight

MARKS:
- Crown over "W" and "M" - William and Mary
- dagger - City of London
- Founders Company guild ewer

MONARCH: William & Mary ORIGIN: English

ILLUSTRATIONS:

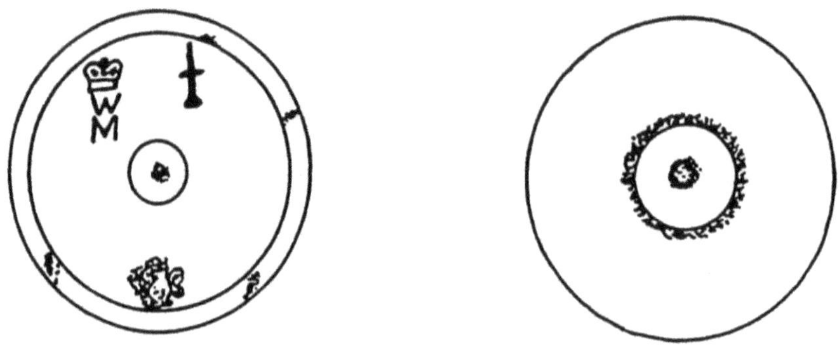


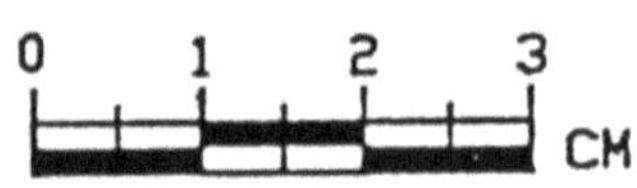


DESCRIPTION: This is a small disc weight which has a dark patina. The markings on the weight are very clear.

DIMENSIONS: HEIGHT (cm): .45
DIAMETER/TOP(cm): 3.18
DIAMETER/BASE(cm): 2.95

WEIGHT IN KILOS: .25 WEIGHT IN POUNDS: .529

PROBABLE WEIGHT: 8 oz A. VARIANCE: +5.8%

PARALLELS: markings on this weight are unique

MATERIALS: brass - high in copper

COMMENTS: The weight has deep score lines on both the top rim and bottom of the weight.

CATALOG NUMBER: 64 ARTIFACT NUMBER: NS.2A9.1b

EXCAVATION: Priddy

WEIGHT TYPE: disc weight

MARKS:
- dagger - City of London
- Founders Company of London guild Ewer
- Crown over "C" - King Charles

ORIGIN: English

ILLUSTRATION:

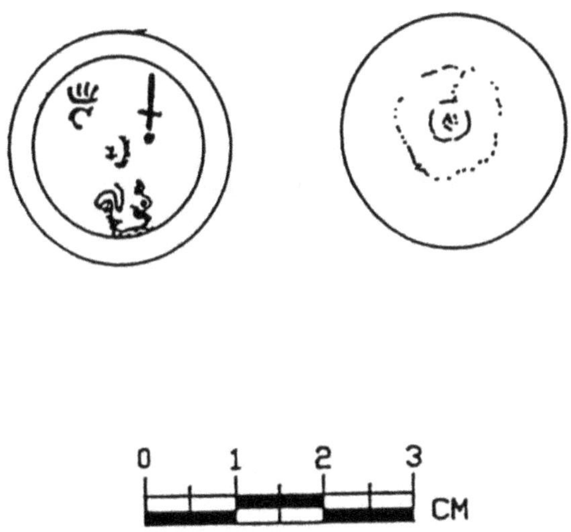


DESCRIPTION: This weight is in excellent condition and appears to fit inside the upper rim of NS.2. A7A.1d. The disc was adjusted to standard by means of a turned groove on the bottom surface of the weight. Final adjustment was made by drilling a small hole to remove some extra metal.

DIMENSIONS: HEIGHT (cm): .5
DIAMETER/TOP (cm): 3.42

WEIGHT IN KILOS: .264 WEIGHT IN POUNDS: .582

PROBABLE WEIGHT: 8 oz A. VARIANCE: +16.4%

PARALLELS: This weight is similar in form to several other weights from the collection.

MATERIALS: bronze

COMMENTS: The workmanship on this weight is excellent - probably part of a set including weight NS.2. A7A.1d.

CATALOG NUMBER: 65 ARTIFACT NUMBER: NS.2.A9.1b

EXCAVATION: Priddy

WEIGHT TYPE: disc weight

MARKS: - Roman Numeral IIII or 4 hatch marks on the top rim

ORIGIN: English

ILLUSTRATION:

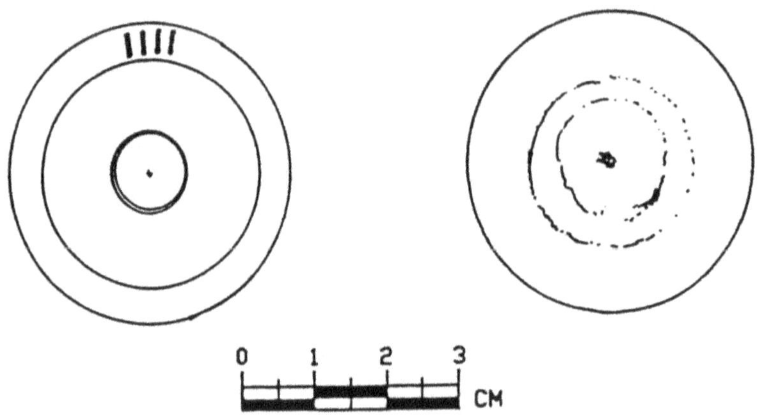


DESCRIPTION: The single unusual feature of this weight is it's high top rim which has created a deep recession in the top surface of the weight. There are two incised circles on the top surface of the weight. The weight was adjusted to standard by carving a circular ring in the bottom surface of the weight. The four hatch marks on the rim of the weight are probably weight designation marks.

DIMENSIONS: HEIGHT (cm): .97
DIAMETER (cm): 4.426

WEIGHT IN KILOS: .117 WEIGHT IN POUNDS: ..258

PROBABLE WEIGHT: 4 oz A. VARIANCE: +3.2%

PARALLELS: This weight has the deepest rim cut of any of our weights.

MATERIALS: bronze

COMMENTS: none

CATALOG NUMBER: 66 ARTIFACT NUMBER: PR85 1024-7

EXCAVATION: Hamilton

WEIGHT TYPE: disc weight

MARKS:
- Crown over "C" - Charles I or II
- dagger - City of London
- Founders Company guild ewer
- A - avoirdupois standard

MONARCH: 17nth century ORIGIN: England

ILLUSTRATIONS:

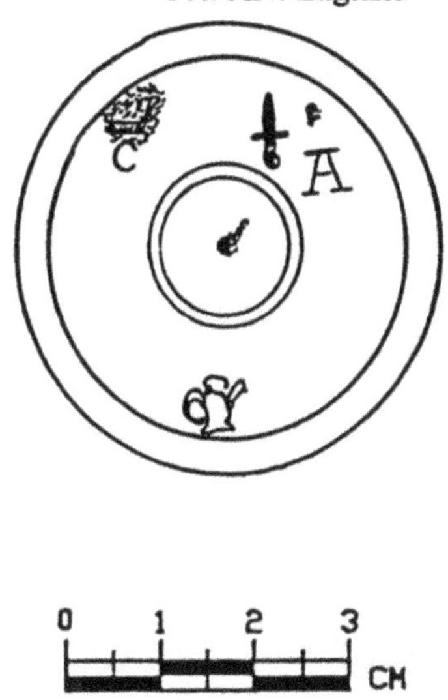


DESCRIPTION: PR85 1024-7 is a small disc weight (4.15 cm dia.) with the handle of the Founders Company guild ewer reversed (on the left) as compared to several other disc weights including NS.2.A7a.1d, with the handle on the right.

DIMENSIONS: HEIGHT (cm): .64
DIAMETER/BASE (cm): 4.15

WEIGHT IN KILOS: .05 WEIGHT IN POUNDS: .11

PROBABLE WEIGHT: 2 oz A. VARIANCE: -12%

PARALLELS: The weight has similar marks as those found on D95-3.

MATERIALS: bronze

COMMENTS: The crown over "C" is partially obscured.

CATALOG NUMBER: 67 ARTIFACT NUMBER: PR85 1024-7.5

EXCAVATION: Hamilton

WEIGHT TYPE: disc weight

MARKS:
- "C" - Charles I or II
- dagger - City of London
- Founders Company guild ewer with the handle on the left
- A - avoirdupois standard

ORIGIN: English

ILLUSTRATIONS:

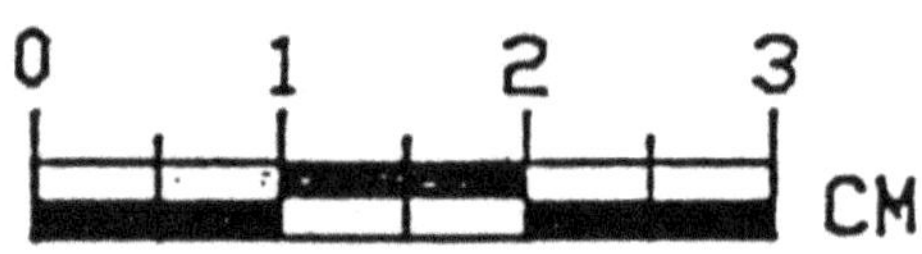


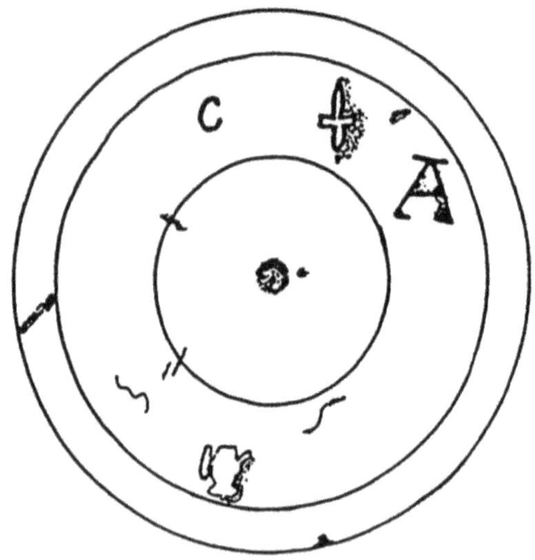


DESCRIPTION: The regis mark differs from most of the other marks recorded from the site since the "C" on this weight is different in that no crown or orb is included over the initial.

DIMENSIONS: HEIGHT (cm): 1.55
DIAMETER/TOP (cm): 6.47
DIAMETER/BASE (cm): 6.17

WEIGHT IN KILOS: .198 WEIGHT IN POUNDS: .44

PROBABLE WEIGHT: 8 oz A. VARIANCE: -12%

PARALLELS: No other weight recovered from the site has a single initial monarch's mark without either a crown or orb.

MATERIALS: bronze

COMMENTS: This weight appears to be darker than most other disc weights and possibly contains a higher percentage of copper.

CATALOG NUMBER: 68 ARTIFACT NUMBER: PR85 1026-7

EXCAVATION: Hamilton

WEIGHT TYPE: disc weight

MARKS: none

ILLUSTRATION:

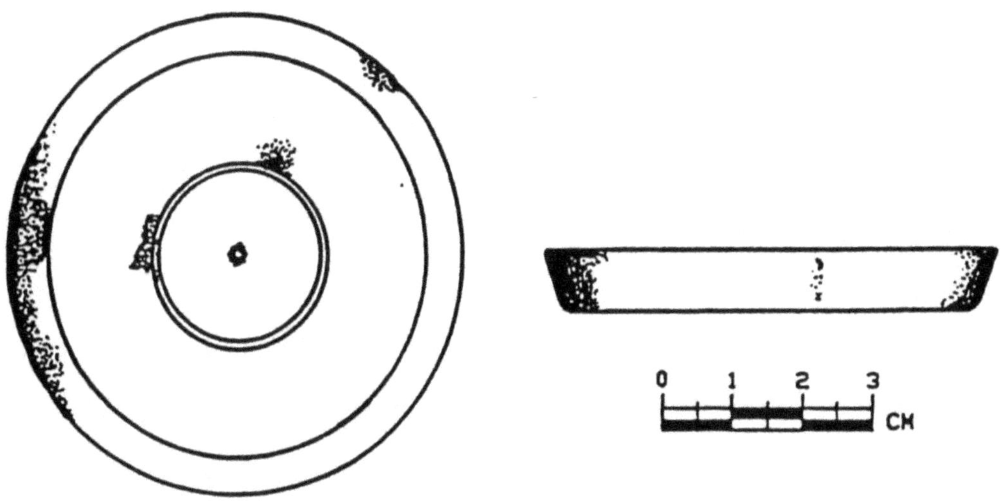


DESCRIPTION: This disc weight is very plain in appearance with a raised outer edge and rounded corners. On the upper surface of the weight are two inscribed circles.

DIMENSIONS: HEIGHT (cm): 1.85
DIAMETER/BASE (cm): 6.549

WEIGHT IN KILOS:.347 WEIGHT IN POUNDS: .765

PROBABLE WEIGHT: 12 oz A. VARIANCE: +2%

PARALLELS: This style weight is very common and several other of the disc weights have the same inscribed lines on their upper surface - its mass however, seems like an unlikely increment.

MATERIALS: bronze

COMMENTS: none

CATALOG NUMBER: 69 ARTIFACT NUMBER: PR85 1034-8

EXCAVATION: Hamilton

WEIGHT TYPE: disc weight

MARKS: none

ILLUSTRATION:

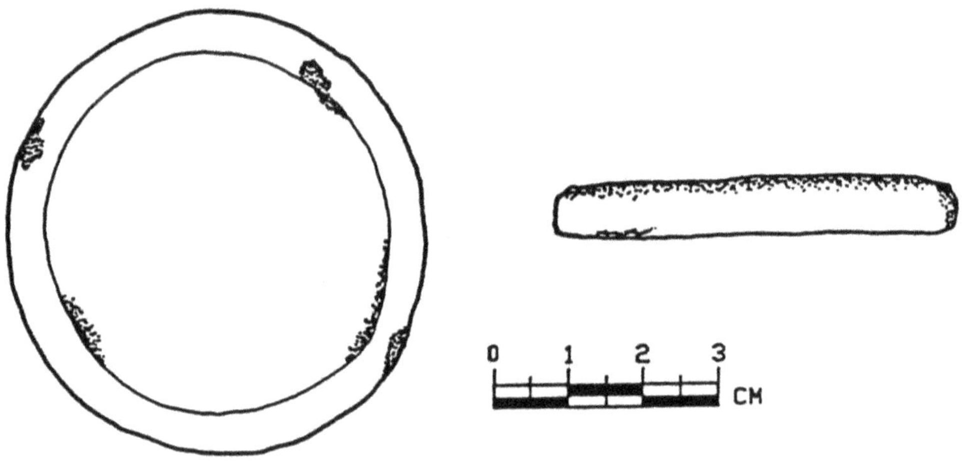


DESCRIPTION: This weight is unique in that there are no marks and inscribed circles on the top surface.

DIMENSIONS: HEIGHT (cm): .7
DIAMETER/BASE (cm): 5.6

WEIGHT IN KILOS: .098 WEIGHT IN POUNDS: .22

PROBABLE WEIGHT: 4 oz A. VARIANCE: -12%

PARALLELS: This weight is similar in shape and style to PR85 1034-6.

MATERIALS: bronze

COMMENTS: none

CATALOG NUMBER: 70 ARTIFACT NUMBER: PR85 1035-6

EXCAVATION: Hamilton

WEIGHT TYPE: disc weight

MARKS: - orb
- 2 daggers

ILLUSTRATION:

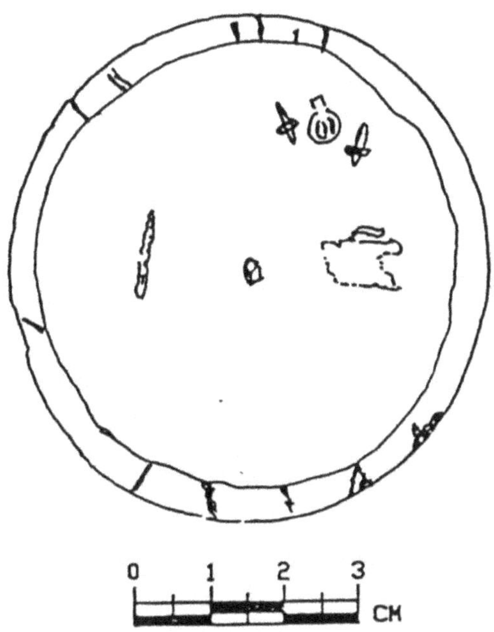


DESCRIPTION: The top surface of the weight is badly pitted.

DIMENSIONS: HEIGHT (cm): 2.3
DIAMETER/BASE (cm): 8.24

WEIGHT IN KILOS: .874 WEIGHT IN POUNDS: 1.93

PROBABLE WEIGHT: 2 lb A. VARIANCE: -3.5%

PARALLELS: This weight is very similar to PR85 1026-8 which also has circular mark indentations in it's upper surface.

MATERIALS: bronze

COMMENTS: The rim of this weight is badly worn - four inscribed lines on the edge may indicate that the weight was part of a larger set.

CATALOG NUMBER: 71 ARTIFACT NUMBER: PR 85 1053-5

EXCAVATION: Hamilton

WEIGHT TYPE: disc weight

MARKS: none

ILLUSTRATION:

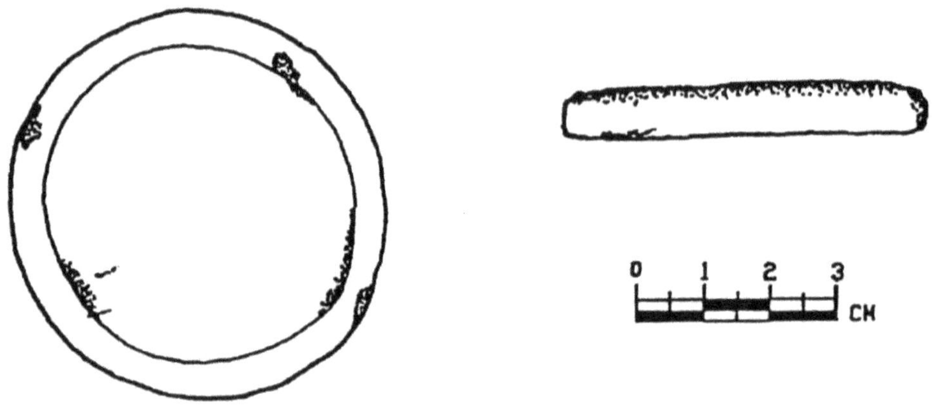


DESCRIPTION: This weight is unique in that it has no markings at all.

DIMENSIONS: HEIGHT (cm): 1.26
DIAMETER/BASE (cm): 8.836

WEIGHT IN KILOS: .195 WEIGHT IN POUNDS: .43

PROBABLE WEIGHT: 8 OZ a. VARIANCE: -14%

MATERIALS: bronze

COMMENTS: none

CATALOG NUMBER: 72

ARTIFACT NUMBER: PR90 1053-6

EXCAVATION: Hamilton

WEIGHT TYPE: disc weight

MARKS: none

ORIGIN: unknown

ILLUSTRATIONS:

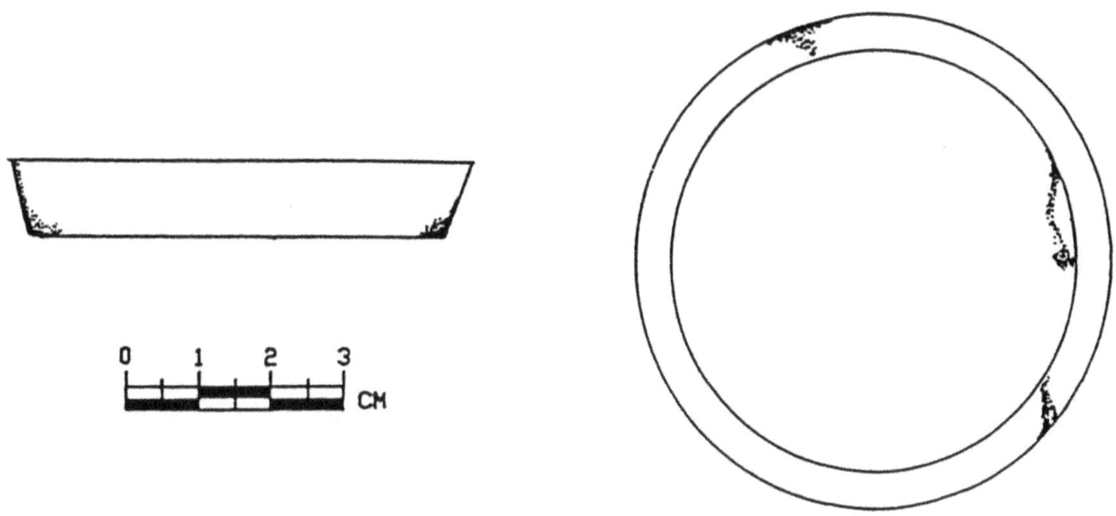


DESCRIPTION: This weight is very plain in design with no parks or ciphers on it.

DIMENSIONS: HEIGHT (cm): 1.0
DIAMETER/BASE (cm): 6.5

WEIGHT IN KILOS: .195

WEIGHT IN POUNDS: .43

PROBABLE WEIGHT: 8 oz A.

VARIANCE: -14%

PARALLELS: none

MATERIALS: bronze

COMMENTS: This weight is very plain in form.

CATALOG NUMBER: 73

ARTIFACT: NUMBER: NP

EXCAVATION: Marx

WEIGHT TYPE: single nested weight

ORIGIN: Germany

ILLUSTRATION:

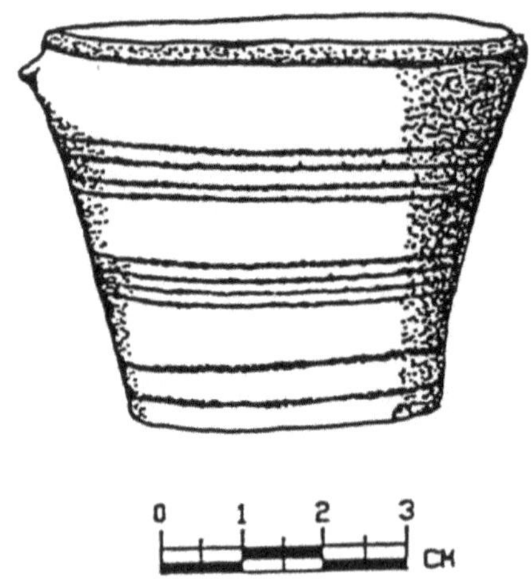


DESCRIPTION: This single nested weight is very similar to the Monte Christi Wreck weights in it's design and engravings. There are signs that bronze was removed from the bottom of this weight although not to the degree as is evident in the Monte Christi weights.

DIMENSIONS: HEIGHT (cm): 2.60
DIAMETER/TOP (cm): 2.35

WEIGHT IN KILOS: .11

WEIGHT IN POUNDS: .243

PROBABLE WEIGHT: 4 oz T.

VARIANCE: -2.8%

MATERIALS: BRONZE

COMMENTS: none

CATALOG NUMBER: 74 ARTIFACT NUMBER: N.P.

EXCAVATION: Marx

WEIGHT TYPE: nested weight set - possibly with six weight cups - lid missing
MARKS: - 2 Troy marks and 2 lion passant on the bottom of the two largest cups
- the third cup has 2 "AB" marks on the bottom (probably the mark of an assayer)

ORIGIN: English

ILLUSTRATIONS:

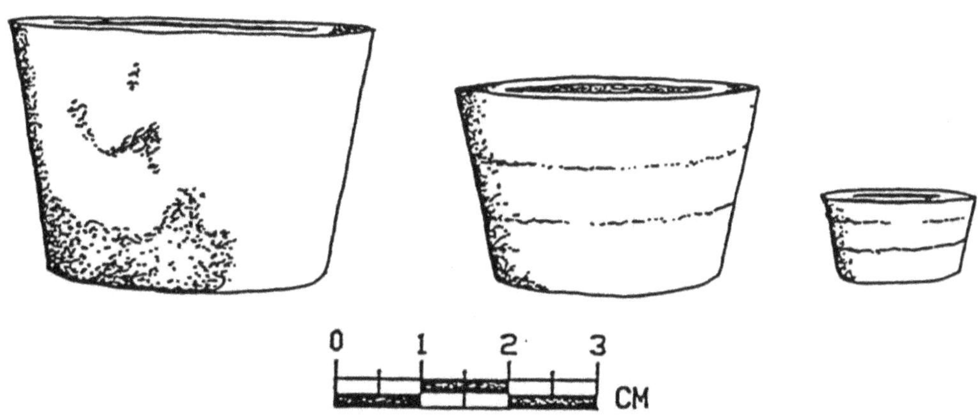


DESCRIPTION: This nested weight set originally consisted of possibly six cups in increments of 32,16,8,4,2 and 1 units as indicated by the Roman numerals located on the upper rims of the weights.

DIMENSIONS:
LARGEST CUP (cm): 9.70 TOP, 6.90 BTM, 8.70 HT - 950.00 gm
- 8 oz T (-14%)

MIDDLE CUP (cm): 8.20 TOP, 6.00 BTM, 6.15 HT - 472.64 gm
- 4 oz T (-14%)

SMALLEST CUP(cm): 6.80 TOP, 4.90 BTM, 4.75 HT - 235.20 gm
- 2 oz T (-12%)

PARALLELS: This nested weight set is similar to PR90 964-7.1, 7.2, 7.3.
MATERIALS: bronze

COMMENTS: possible English

CATALOG NUMBER: 75 ARTIFACT NUMBER: N.P.

EXCAVATION: Marx

WEIGHT TYPE: nested weights - possibly a set of five originally

MARKS: - 2 Troy marks and 2 lion passant on the bottoms of the two remaining largest cups

ORIGIN: English

ILLUSTRATIONS:

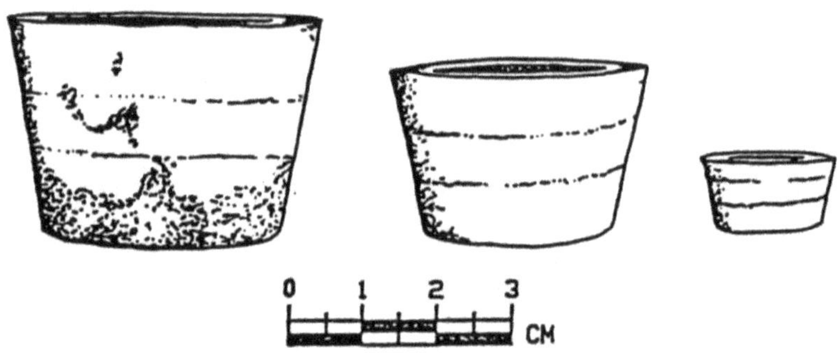


DESCRIPTION: The weight set originally consisted of five cups, possibly including a disc weight, but only three cups were recovered. Roman numerals were used to number the cups on the upper rim (IIII-largest, III, II).

DIMENSIONS:
LARGEST CUP (cm):4.10 TOP, 3.00 BTM, 2.70 HT. - 118.40 gm - 4 oz T (+4%)

MIDDLE CUP (cm):3.20 TOP, 2.40 BTM, 2.00 HT. - 61.20 gm - 2 oz T (+3.76%)

SMALL CUP (cm):1.60 TOP, 1.40 BTM, .9 HT. - .09 gm - 1 oz T (+11%)

PARALLELS: These weights cups were manufactured in the most common style of German craftsmen.

MATERIALS: bronze

COMMENTS: This set could be either German or English. During the late 1600's Nuremberg was manufacturing weights for many countries in different weight standards.

CATALOG NUMBER: 76 ARTIFACT NUMBER: Institute of Jamaica

EXCAVATION: unknown

WEIGHT TYPE: nested weight set (3 cups recovered) - the outer cup has a lid attached

MARKS:
- "LXIIII"
- 2 lion passant
- 2 Troy marks

ORIGIN: England

ILLUSTRATIONS:

DESCRIPTION: The outer cup and lid of the set weighed 5 pounds (2.01 l.). If the outer weight cup weighs the total of the remaining weight cups, this was a ten pound nested weight set.

DIMENSIONS:

| | | | |
|---|---|---|---|
| OUTER CUP | (cm):9.40 TOP, 4.78 BTM, 8.75 HT - 2.01 k | - 5 lb T | (0.0%) |
| CUP 2 | (cm):8.30 TOP, 6.08 BTM, 6.20 HT - 1.08 k | - 2 lb 6 oz T | (0.0%) |
| CUP 3 | (cm):6.80 TOP, 4.95 BTM, 2.70 HT - .54 k | - 2 lb T | |
| CUP 4 | na | - 6 oz T | |

MATERIALS: bronze

COMMENTS: The design of this nested weight set is identical to an English set located in the Streeter Collection (Danforth 1988, p.100).

CATALOG NUMBER: 77 ARTIFACT NUMBER: U.M.L.I.S.T.

EXCAVATION: unknown

WEIGHT TYPE: lid only from a nested weight set

MARKS: (V)III - possibly an eight ounce weight set.

ORIGIN: English

ILLUSTRATIONS:

DESCRIPTION: Part of the marking on the top of the lid has been obscured by deterioration of the metal. This lid was part of a small weight set probably incorporating ounce units of 8,4,2,1, .5 and a .5 ounce disc totaling 16 ounces. While this is a likely combination of weights based on the lid markings and the size of the lid, it is speculative since only the lid was recovered.

DIMENSIONS: HEIGHT (cm): .35
DIAMETER/TOP(cm): 4.00

PARALLELS: Similar to the lid design of an English nested weight set using the Troy standard (Danforth 1988, p.100).

MATERIALS: bronze

COMMENTS: none

CATALOG NUMBER: 78 ARTIFACT NUMBER: SP.1 S6

EXCAVATION: St. Peter's Church, Mayes

WEIGHT TYPE: single weight cup

MARKS: none

ORIGIN: unknown

ILLUSTRATIONS:

DESCRIPTION: This weight cup was found at St. Peter's Church excavation, layer 1, structure #6. This single weight cup was in good condition with very little corrosion. There were no marks on the weight cup.

DIMENSIONS: HEIGHT (cm): .60
DIAMETER/TOP(cm): 1.30
DIAMETER/BASE(cm): 1.20

WEIGHT IN KILOS: .03 WEIGHT IN POUNDS: .066

PROBABLE WEIGHT: 1 oz T VARIANCE: +1.53%

PARALLELS: The weight cup is similar to PR90 941.

MATERIALS: bronze

COMMENTS: none

CATALOG NUMBER: 79 ARTIFACT NUMBER: PR89 897-7.1

EXCAVATION: Hamilton

WEIGHT TYPE: single nested eight

MARKS: none

ILLUSTRATION:

DIMENSIONS: The style and design of this weight cup is identical to other nested weights which were manufactured in Nuremberg. On the inner bottom surface of the weight is an inscribed "4" mark.

DIMENSIONS: HEIGHT (cm): 2.85<br>
DIAMETER/BASE (cm): 2.40<br>
DIAMETER/TOP (cm): 3.20

WEIGHT IN GRAMS: 54 WEIGHT IN POUNDS: .119

PROBABLE WEIGHT: 2 oz T VARIANCE: -4.5%

PARALLELS: This weight cup is very similar to PR90 964-7.1

MATERIALS: bronze

COMMENTS: This weight cup is probably German in origin. Similar to PR90 964-7.1, this weight has incised lines on its inner and outer surfaces.

CATALOG NUMBER: 80 ARTIFACT NUMBER: PR90 941

EXCAVATION: Hamilton

WEIGHT TYPE: single nested weight cup

MARKS: none

ORIGIN: unknown

ILLUSTRATIONS:

DESCRIPTION: This single weight cup was badly corroded when it was recovered and no marks were detected on any surfaces.

DIMENSIONS: HEIGHT (cm): .90
DIAMETER/TOP (cm): 2.54

WEIGHT IN GRAMS 12.5 WEIGHT IN POUNDS: .03

PROBABLE WEIGHT: .5 oz T VARIANCE: -12%

PARALLELS: This weight cup is similar to SP.1 S6.

MATERIALS: bronze

COMMENTS: This weight cup was probably lost from a more complete set. A single weight such as this one may have been used for a special purpose such as weighing precious metals or for apothecary measurement, independent of a complete set. Based on this single cup, origin can not be determined.

CATALOG NUMBER: 81 ARTIFACT NUMBER: PR89 964-7.1, 7.2, 7.3

EXCAVATION: Hamilton

WEIGHT TYPE: nested weight set - three cups recovered

MARKS: - rims marked with Roman numerals XVI, VIII, IIII
- 2 Troy marks and 2 lion passant on bottom of each cup

ORIGIN: English

ILLUSTRATIONS:

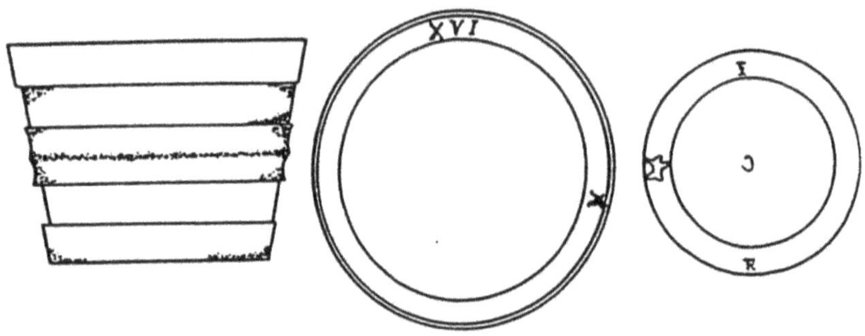


DESCRIPTION: The outer cup design is more ornate than those manufactured in England during the mid 1600's. Three cups were recovered, all with Troy marks and lion passant on the bottom rims and Roman numerals on the top rims.

DIMENSIONS:
LARGEST CUP (cm): 6.50 TOP, 4.95 BTM, 4.75 HT - .48 k - 1 lb T

MIDDLE CUP (cm): 5.4O TOP, 3.8O BTM, 3.60 HT - .24 k - 8 oz T

SMALLEST CUP (cm): 4.30 TOP, 3.18 BTM, 2.65 HT - .12 k - 4 oz T

PARALLELS: The design of the outer cup is similar to the Monte Christi Wreck weights. Similar outer cup designs are included in the Streeter Collection (Danforth 1988, p.53 & 64).

MATERIALS: bronze

COMMENTS: Because all three cups have the pronounced kick or indentation on their bottoms which German weights generally do not have, as well as the lions passant and Troy stamps, it is most probable that these weight cups are English in origin.

CATALOG: 83 ARTIFACT NUMBER: PR NP

EXCAVATION: surface find, Richard McClure

WEIGHT TYPE: Dram Weight

MARKS: 2 dot/circles and 1 undistinguishable mark - similar marks are evident on both sides of the weight

ILLUSTRATION:

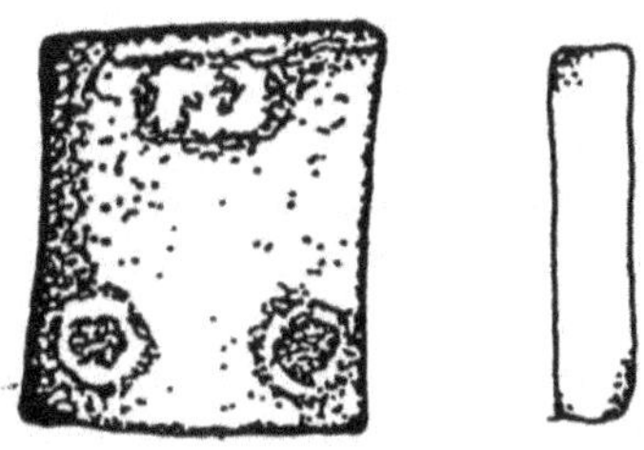

DESCRIPTION: This weight was not available for analysis but photographs of the weight suggest that the weight showed signs of heavy use-wear. Because this weight was found by the hospital wall in loose dirt, much of the heavy scoring on the weight is probably due to its provenance. The fine feature of the weight suggest that this weight was a precisely manufactured apothecary weight.

DIMENSIONS: LENGTH (cm): 1.00
WIDTH (cm): 1.00
THICKNESS (cm): .20

WEIGHT IN KILOS: NA WEIGHT IN POUNDS: NA

PARALLELS: similar to NS.2.A2b.1.

MATERIALS: bronze

COMMENTS: none

CATALOG NUMBER: 84 ARTIFACT NUMBER: NS.2.A2b.1.

EXCAVATION: Priddy

WEIGHT TYPE: precious metals/ apothecary weight

MARKS: - Roman numeral I III
- 2 obscured marks

ILLUSTRATION:

DESCRIPTION: This is a small round weight with a raised dome handle. The profile of the weight indicates that the base is slightly smaller than the top surface of the weight.

DIMENSIONS: HEIGHT (cm): .746
DIAMETER/BASE (cm): 1.2
DIAMETER/TOP (cm): 1.42

WEIGHT IN KILOS: NA WEIGHT IN POUNDS: NA

MATERIALS: bronze

COMMENTS: This weight was commonly used in apothecaries and for the measurement of precious metals.

CATALOG NUMBER: 85 ARTIFACT NUMBER: NS.2.A2b.1.

EXCAVATION: Priddy

WEIGHT TYPE: Dram Weight

MARKS: "WR" mark on both surfaces of the weight

ORIGIN: Commonly used throughout Europe

ILLUSTRATION:

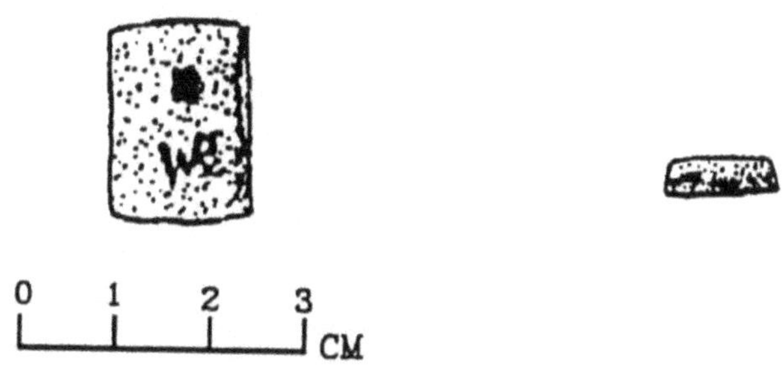


DESCRIPTION: This weight is square in shape with a slightly beveled top edge. Both top and bottom surfaces of the weight have a "WR" inscription inscribed on them.

DIMENSIONS: HEIGHT (CM): .4
WIDTH (CM): 1.65
LENGTH (CM): 1.4

WEIGHT IN KILOS: NA WEIGHT IN POUNDS: NA

MATERIALS: bronze

COMMENTS: This is a dram weight, similar to the one recovered during Marx's excavations but in better condition.

CATALOG NUMBER: 86 ARTIFACT NUMBER: MW-91

EXCAVATION: Marx

WEIGHT TYPE: precious metal weight (not available for analysis)
MARKS: - not able to determine what the mark is based on the sketch located in the notes

ORIGIN: common throughout Europe

ILLUSTRATIONS:

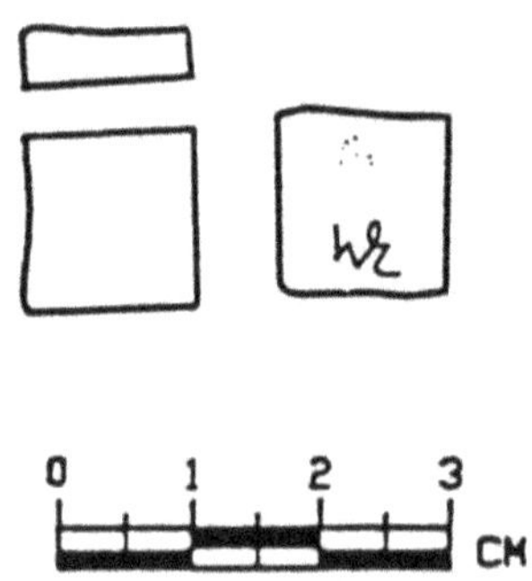


DESCRIPTION: The weight is one half inch square and approximately three-sixteenths of an inch thick. No reference was located in any of Marx's notes as to the weight of the artifact.

DIMENSIONS: 1.2 cm (.5 in.) square
THICKNESS (cm):.45

WEIGHT IN KILOS: N.A. WEIGHT IN POUNDS: N.A.

PARALLELS: Similar weights are illustrated in Kisch 1965, p.53.

MATERIALS: bronze

COMMENTS: Weights such as this were commonly used for weighing precious metals such as gold and silver, as well as coins.

CATALOG NUMBER: 87

ARTIFACT NUMBER: SP-91

EXCAVATION: Marx

WEIGHT TYPE: scale pan

MARKS: Crown over I♥H

ORIGIN: unknown

ILLUSTRATIONS:

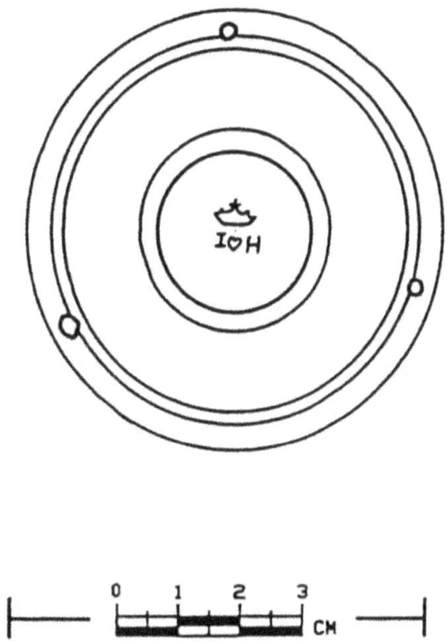


DESCRIPTION: Three holes equally spaced around the circumference of the scale pan indicate that the pan was probably supported by three small chains with hook ends or by cords secured through them. The pan has decoratively inscribed circles with a Crown and I ♥ H mark located in the middle of the pan.

DIMENSIONS: THICKNESS (cm): .15
DIAMETER/BASE (cm): 6.65

PARALLELS: none

MATERIALS: bronze

COMMENTS: The scale pan is small enough to suggest that it was probably used for precious metals or apothecary measuring.

CATALOG NUMBER: 88 ARTIFACT NUMBER: PR85 1026-14

EXCAVATION: Hamilton

WEIGHT TYPE: scale pan

ORIGIN: unknown

ILLUSTRATION:

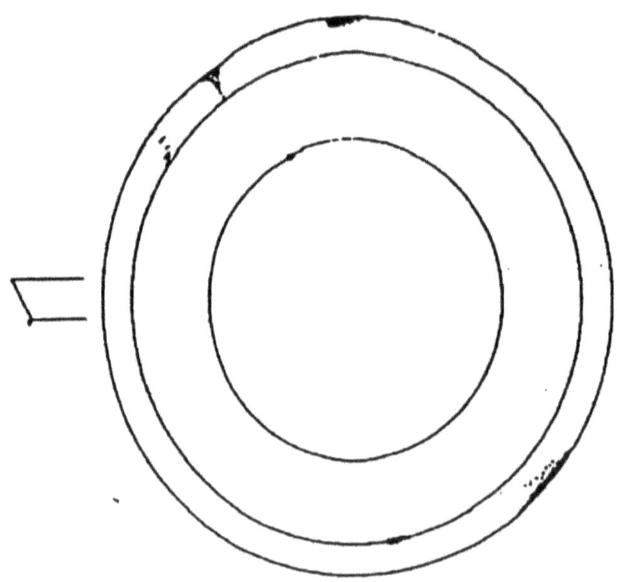

DESCRIPTION: The pan is brass and was attached by small chains at three points along its rim.

DIMENSIONS: DEPTH (cm): 1.5
DIAMETER (cm): 20.2

PARALLELS: This type of scale pan is very common although this pan has a small foot ring which SP-91 does not have.

MATERIALS: brass

COMMENTS: The workmanship of this scale pan is much finer than SP-91 although the artifact was slightly bent when recovered.

CATALOG NUMBER: 89 ARTIFACT NUMBER: PR89 712-9

EXCAVATION: Hamilton

WEIGHT TYPE: possibly a scale pan

MARKS: none

ILLUSTRATION:

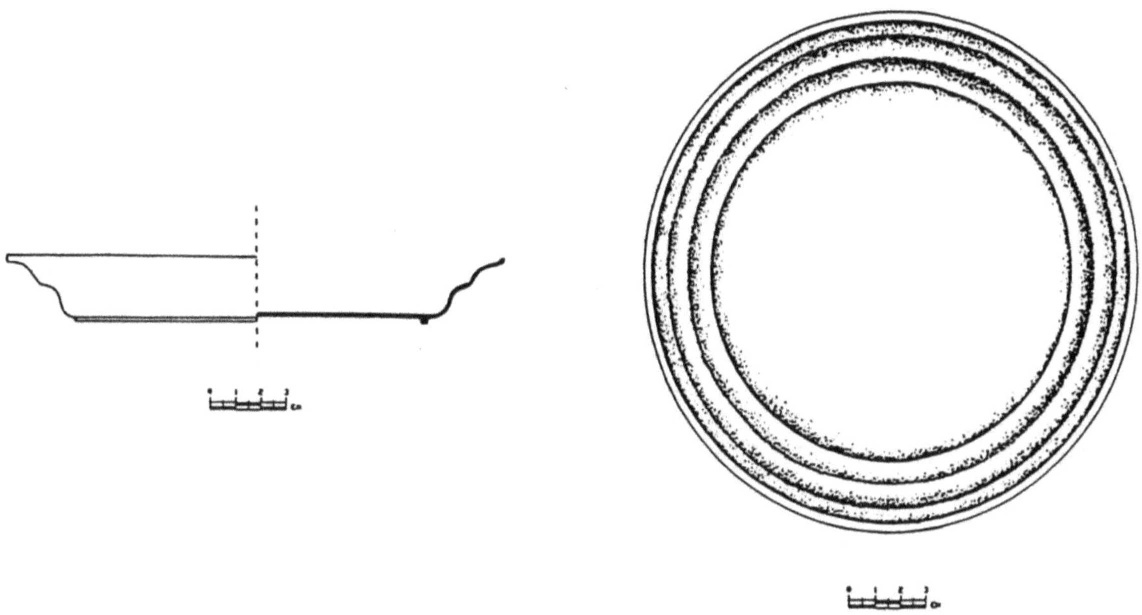

DESCRIPTION: This pan was badly deteriorated and was partially reconstructed by making casts of the concretion. The pan has a small foot ring (indicated in X-rays) and sharply angled sides and rim.

DIMENSIONS: DEPTH (cm): 2.1
DIAMETER/RIM (cm): 19.4

PARALLELS: Very similar to PR85 1026-14 in design and physical dimensions.

MATERIALS: iron

COMMENTS: Based on similarity of design, it is possible that this is a scale pan.

CATALOG NUMBER: 90 ARTIFACT NUMBER: Marx N.P.

EXCAVATION: Marx

WEIGHT TYPE: large balance beam

MARKS: none

ORIGIN: Europe

ILLUSTRATIONS:

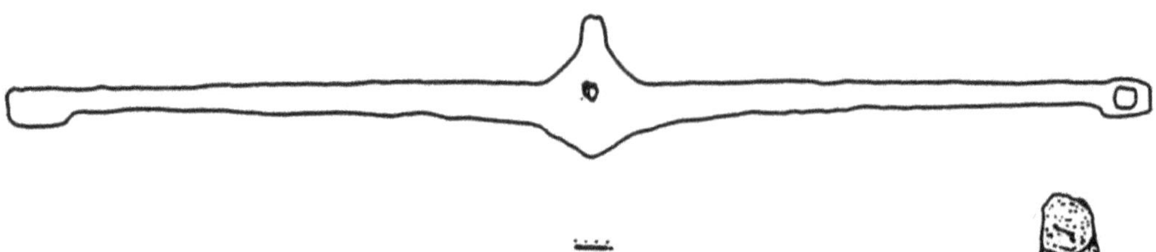

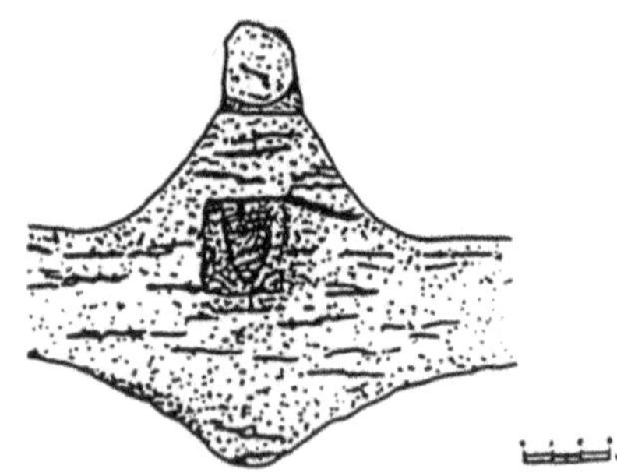

DESCRIPTION: This is the largest of the balance beams recovered from the Port Royal excavations. This is a simple balance beam consisting of two centrally located fulcrum blades (upon which the arm pivoted) and short attachment arms (one at each end of the beam), to which pans or hooks could be attached.

DIMENSIONS: HEIGHT (cm): 15.30
CONSERVED LENGTH: 124.00 cm.

PARALLELS: This is a very common style of beam arm.

MATERIALS: iron

COMMENTS: This balance beam is capable of weighing very heavy objects. The knife edge fulcrum was hammered into place using squared stock that was first heated red hot. The attachment rods to which a balance pan or hook would have been fastened were attached to the main body of the balance beam in a similar fashion. The indicator arm was fastened to the beam arm using a single large rivet.

CATALOG NUMBER: 91 ARTIFACT NUMBER: PR89 897-9.11

EXCAVATION: Hamilton

WEIGHT TYPE: small balance beam arm with pointer

ILLUSTRATION:

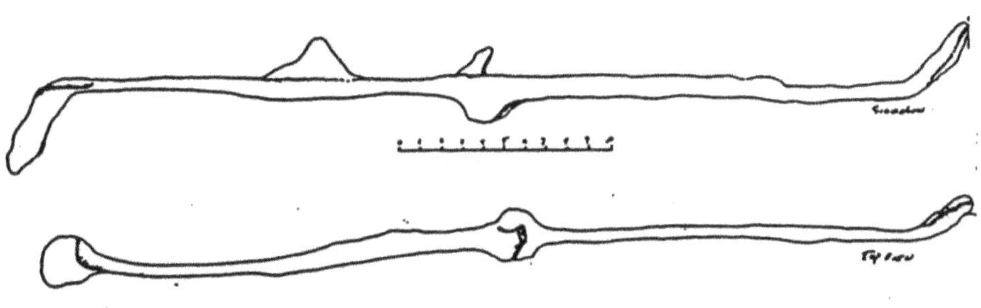

DESCRIPTION: This beam arm was part of a large concretion found in the hearth area of Building Five. The beam is in very deteriorated condition and has been reconstructed using Hysol casts. The beam has an eighteen inch span from end to end.

DIMENSIONS: LENGTH (cm): 18.6
DIAMETER (cm): .8

MATERIALS: iron

PARALLELS: This is a very common scale type.

COMMENTS: The ends of the beam arm are very deteriorated but the appear to have had a hole running through them to accommodate a hook holding the cords of suspended scale pans.

CATALOG NUMBER: 92 ARTIFACT NUMBER: PR85 1022-5

EXCAVATION: Hamilton

WEIGHT TYPE: balance beam

MARKS: none

ORIGIN: Europe

ILLUSTRATIONS:

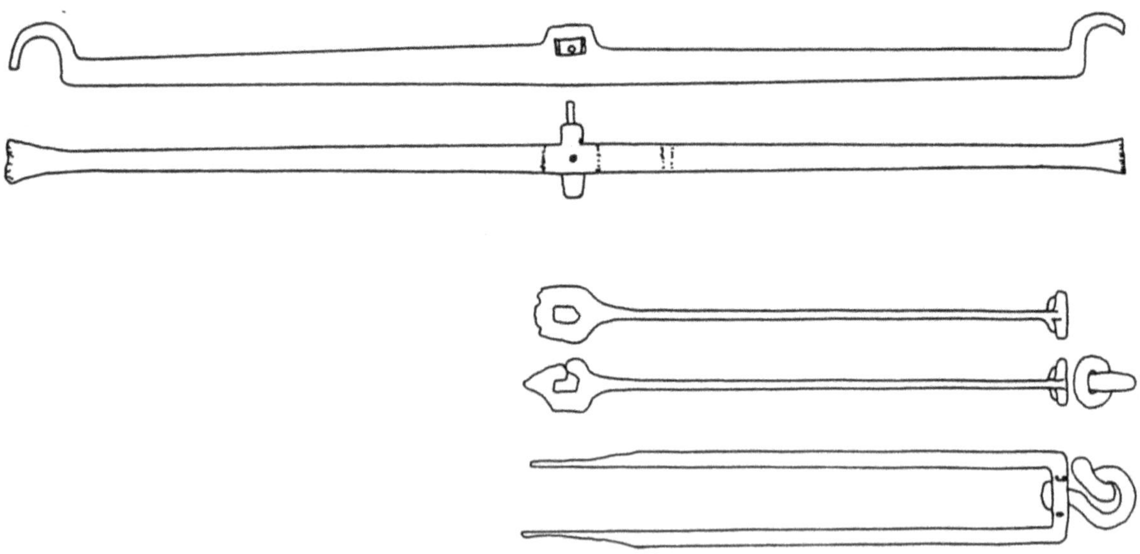

DESCRIPTION: The beam is in four pieces consisting of the arm itself, a portion of the indicator needle, as well as two support arms through which the knife edge located centrally on the beam arm would have passed. The arms were equidistant from the knife edge fulcrum point of the beam.

DIMENSIONS: HEIGHT (cm): 3.00
LENGTH (cm): 57.50

PARALLELS: Similar to PR85 1025-7.

MATERIALS: iron

COMMENTS: This style of balance beam arm is very common and the style was used for small and large beams alike.

CATALOG NUMBER: 93 ARTIFACT NUMBER: PR85 1025-7

EXCAVATION: Hamilton

WEIGHT TYPE: balance beam

MARKS: none

ORIGIN: Europe

ILLUSTRATIONS:

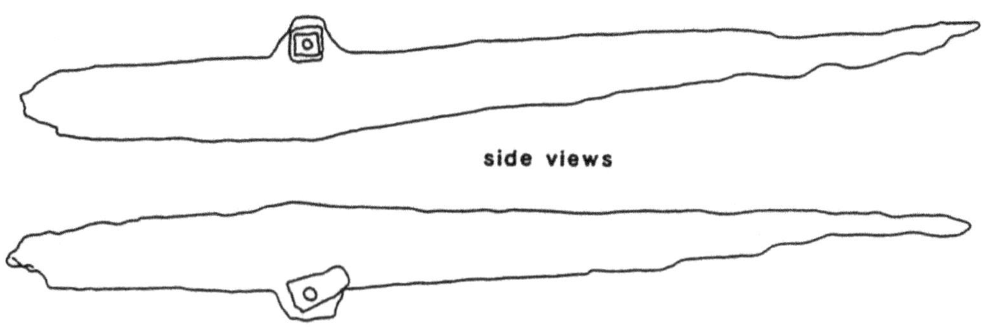

DESCRIPTION: Like PR85 1022-5, this balance beam has suffered great deterioration, largely because it was made of iron. The arms of the beam were equidistant from the centrally located fulcrum of the beam and the size of the central section of the arm suggests that this beam was of heavier construction than PR84 1022-5.

DIMENSIONS:

THICKNESS AT CENTER (cm): 5.60
CONSERVED LENGTH (cm): 98.00

PARALLELS: Similar to PR85 1022-5.

MATERIALS: iron

COMMENTS: Knife edges are fashioned from squared blocks that are press fitted into the body of the beam and allowed to contract, forming a very strong joint with the main body of the beam.

CATALOG NUMBER: 94 ARTIFACT NUMBER: PR85 1025-1

EXCAVATION: Hamilton

WEIGHT TYPE: weight adjuster

MARKS: - One squared hole in the center of the weight

ILLUSTRATION:

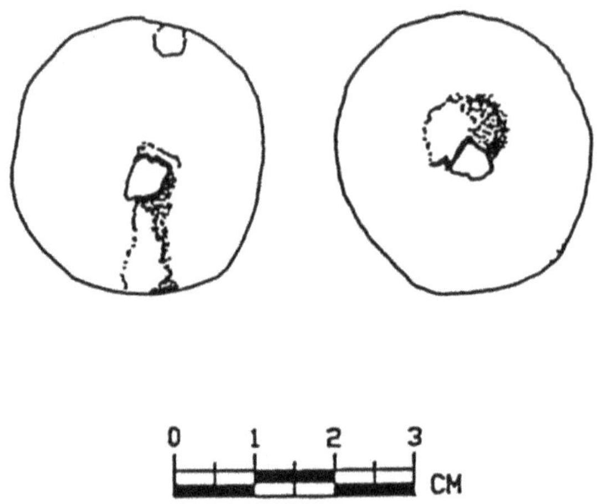


DESCRIPTION: This weight adjuster has one squared hole in it's center as well as a sharp indentation along the edge. The adjuster is concave in shape and was probably nailed onto the rounded top surface of a dome weight.

DIMENSIONS: HEIGHT (cm): .1824
DIAMETER/TOP (cm): 3.873

WEIGHT IN KILOS: NA WEIGHT IN POUNDS: NA

PARALLELS: This adjuster is similar to the others from the site.

MATERIALS: lead

COMMENTS: Iron oxide on the concave surface of this weight indicates that it was fastened to another weight using an iron nail.

CATALOG NUMBER: 95 ARTIFACT NUMBER: PR89 676-5

EXCAVATION: Hamilton

WEIGHT TYPE: Weight adjuster

MARKS: four holes

ILLUSTRATIONS:

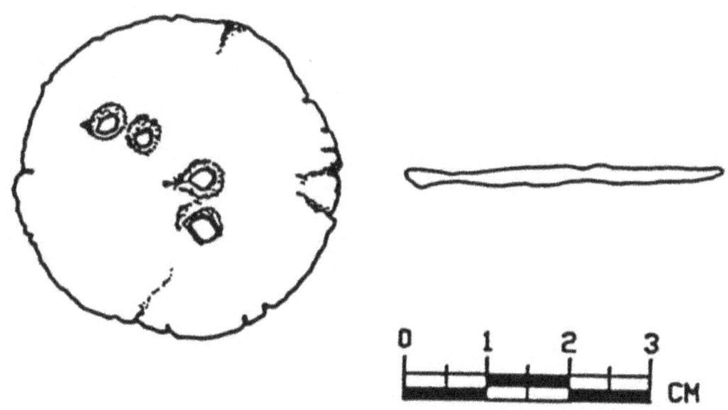


DESCRIPTION: The four squared holes located on the flat surface of this weight adjuster indicated that it was probably nailed onto the body of a larger weight. The edges of the lead are crazed and cracked, indicating that the lead was probably hammered into shape from a small cast bun of lead.

DIMENSIONS: HEIGHT (cm): .1796
DIAMETER/BASE (cm): 3.7920

WEIGHT IN GRAMS: 16.27

PARALLELS: This weight adjuster is very similar to several other weights including PR85 1024-8.0, 8.1, 8.2, 8.3, 8.4, 8.5 and PR85 1023-5.

MATERIALS: lead

COMMENTS: The average weight of all weight adjusters is 20.6637 grams.

CATALOG NUMBER: 96 ARTIFACT NUMBER:PR85 1024 - 8.0
1024 - 8.1
1024 - 8.2
1024 - 8.3
1024 - 8.4
1024 - 8.5

EXCAVATION: Hamilton

WEIGHT TYPE: small disc weights

MARKS: none

ORIGIN: unknown

ILLUSTRATIONS:

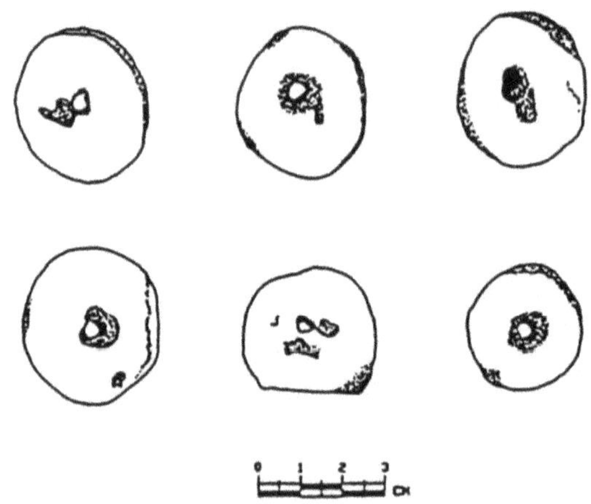

DESCRIPTION: All six weights have holes through their centers and are circular in shape. The edge of weight 1024-8.4 has been trimmed as a means of altering the weight.

DIMENSIONS: HEIGHT (cm):all approximately .5

DIAMETER/BASE (cm): 3.0 average

WEIGHT IN GRAMS :#8.0 - 20.4 - .72 oz
#8.1 - 17.8 - .63 oz
#8.2 - 23.2 - .82 oz
#8.3 - 27.0 - .95 oz
#8.4 - 17.9 - .63 oz
#8.5 - 18.5 - .65 oz

PARALLELS: none

MATERIALS: lead

COMMENTS: The increments between weights are very small and these weights were probably used for pharmaceutical measurement as well as for measuring precious metals. Similar weights are displayed at the Science Museum of London, from the Castereagh Collection.

CATALOG NUMBER: 97

ARTIFACT NUMBER: PR85 1023-5

EXCAVATION: Hamilton

WEIGHT TYPE: weight adjuster

MARKS: none

ILLUSTRATION:

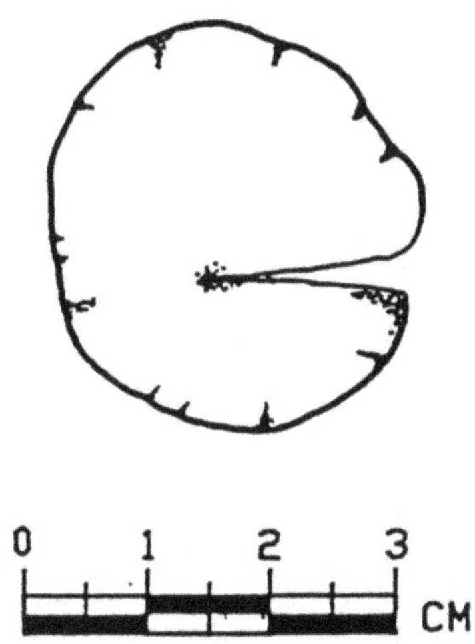


DESCRIPTION: This weight is very similar in shape to PR89 676-5. There is a cut extending from the edge towards the center of the weight, terminating at the edge of the small hole made by a nail fastener.

DIMENSIONS: HEIGHT (cm): .50
DIAMETER/BASE (cm): 2.90

WEIGHT IN KILOS: .023

WEIGHT IN OUNCES: 1.44

PARALLELS: This weight adjuster is also similar to PR 1024-8.0 - 8.5.

MATERIALS: lead

COMMENTS: Weight adjusters were commonly used to fix use-worn weight, to reinstate them to their original standard.

CATALOG NUMBER: 98　　　　ARTIFACT NUMBER: NS.2.A5a.1d

EXCAVATION: Priddy

WEIGHT TYPE: slotted disc weight

ILLUSTRATION:

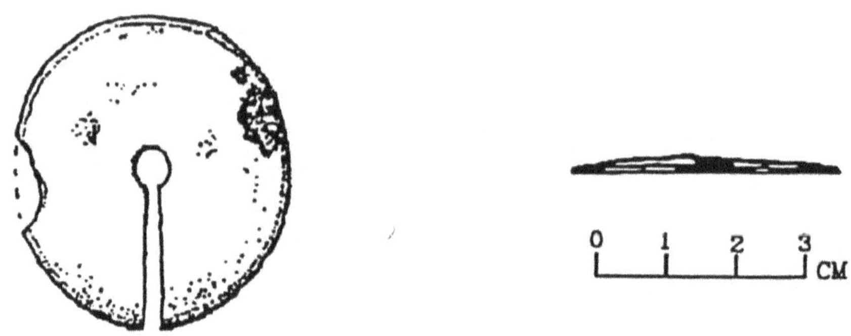


DESCRIPTION: The fine beveled edge of this weight and the precisely drilled central hole indicates that this weight was made by a skilled craftsman. The weight us very thin and a portion of the edge has broken away. Apart from the slot, this weight is small enough that it resembles a lead weight adjuster in design.

DIMENSIONS: HEIGHT (cm): .257
DIAMETER (cm): 3.842

WEIGHT IN KILOS: .0144　　　　WEIGHT IN POUNDS: .5 oz A.

MATERIALS: bronze

COMMENTS: This weight would have been used on a small money scale or beam, and the slot in its edge would have allowed the weight to be centered for accurate measurements.

# Index

## A



## B



## C

**D**



**E**

www.ingramcontent.com/pod-product-compliance
Lightning Source LLC
Chambersburg PA
CBHW042029210726
48526CB00057B/7654